6th Workshop on Infrared Spectroscopy and Microscopy with Accelerator-Based Sources

(WIRMS11)

Journal of Physics: Conference Series Volume 359

Trieste, Italy
4-8 September 2011

ISBN: 978-1-62276-776-2
ISSN: 1742-6588

Printed from e-media with permission by:

Curran Associates, Inc.
57 Morehouse Lane
Red Hook, NY 12571

Printed by Curran Associates, Inc. (2013)

Institute of Physics
Dirac House, Temple Back
Bristol BS1 6BE UK

Phone: 44 1 17 929 7481
Fax: 44 1 17 920 0979

techtracking@iop.org

Additional copies of this publication are available from:

Curran Associates, Inc.
57 Morehouse Lane
Red Hook, NY 12571 USA
Phone: 845-758-0400
Fax: 845-758-2634
Email: curran@proceedings.com
Web: www.proceedings.com

TABLE OF CONTENTS

FACILITIES

MICROSPECTROSCOPY

TIME-RESOLVED SPECTROSCOPIES

EXTREME CONDITIONS

NEAR-FIELD

THZ TECHNIQUES

Journal of Physics: Conference Series **359** (2012) 011001 doi:10.1088/1742-6596/359/1/011001

Preface

This volume of *Journal of Physics: Conference Series* is dedicated to a subset of papers related to the work presented at the 6th edition of the international Workshop on Infrared Spectroscopy and Microscopy with Accelerator-Based Sources (WIRMS), held in Trieste, Italy, September 4–8 2011. Previous editions of the conference were held in Porquerolles (France), Lake Tahoe (USA), Rathen (Germany), Awaji (Japan), and Banff (Canada). This edition was organized and chaired by Stefano Lupi (Roma La Sapienza) and co-chaired by Andrea Perucchi (Elettra), with the support of the Italian Synchrotron Light Laboratory, ELETTRA, which was honored to host the WIRMS workshop in its tenth anniversary.

The 6th WIRMS edition addressed several different topics, ranging from biochemistry to strongly correlated materials, from geology to conservation science, and from forensics to the study of cometary dusts. This edition was attended by 88 participants, including representatives from the infrared scientific programs at synchrotron light sources and free-electron-laser facilities, who enjoyed the stimulating scientific presentations, several detailed discussions, and the beautiful weather and scenery of the Trieste gulf.

Participants came from 16 different nations and four continents, including many young scientists, six of which were supported by the organizers. There were 45 scientific talks divided in 11 sessions: Facilities, Microspectroscopy (I, II, III), Time-Resolved Spectroscopies, Extreme Conditions, Condensed Matter, Near-Field, Imaging and THz Techniques and High-Resolution Spectroscopy. 37 posters were also presented at two very lively evening poster sessions.

We would like to use the opportunity of writing this preface to thank all the participants of the workshop for the very high level of their scientific contribution and for the very friendly atmosphere, which were the most important keys to the success of the workshop. We also wish to thank all the companies (Bruker, Tydex, Spiricon, Allectra, Neaspec, Micos) and institutions (Sincrotrone Trieste, Sapienza-University of Rome, EMS, INFN, CNR-IOM) sponsoring this event. We are indebted to the members of the International Advisory Committee for their valuable help in the difficult choosing of the invited speakers, and for more general advice on the conference organization and structuring. A special acknowledgment goes to the ELETTRA secretary staff, and in particular to Mariella Antonetti, who has been involved with the organization from the very beginning.

The 7th WIRMS conference will be held in Melbourne (Australia). Our best wishes go to the organizers for a successful conference.

Additional information on the conference can still be found on the conference website:
http://www.elettra.trieste.it/events/2011/WIRMS/

A selection of the conference pictures is also available on Flickr:
http://www.flickr.com/photos/sincrotronets/sets/72157627705743677

Guest Editors
Stefano Lupi and Andrea Perucchi

Local Organizing Committee

S Lupi (Chair)	Sapienza University of Rome
A Perucchi (Co-Chair)	Sincrotrone Trieste S.C.p.A.
L Vaccari	Sincrotrone Trieste S.C.p.A.
M Kiskinova	Sincrotrone Trieste S.C.p.A.

International Advisory Committee

J Allen	UCSD (USA)
D Basov	UCSD (USA)
B Billinghurst	CLS (Canada)
G L Carr	NCLS (USA)
P Dumas	SOLEIL (France)
G Edwards	Duke FEL (USA)
G P Gallerano	ENEA-Frascati (Italy)
M Helm	ELBE (Germany)
H-Y Holman	LBNL (USA)
F Keilmann	MPI (Germany)
S-I Kimura	UVSOR (Japan)
Y C Lee	NSRRC (Taiwan)
M Martin	ALS (USA)
Y-L Mathis	ANKA (Germany)
T May	CLS (Canada)
L Miller	NSLS (USA)
U Schade	BESSY II (Germany)
M Tobin	Australian Synchrotron (Australia)
G P Williams	Jefferson Laboratories (USA)

Journal of Physics: Conference Series **359** (2012) 011001 doi:10.1088/1742-6596/359/1/011001

Sponsors

Sincrotrone Trieste S.C.p.A.

'Sapienza' University of Rome

Istituto Nazionale di Fisica Nucleare

CNR-Istituto Officina dei Materiali

European Microscopy Society

Bruker Optics Italia S.r.l.

Tydex

Neaspec

Spiricon

Allectra GmbH

Micos Italia GmbH

Peer review statement

All papers published in this volume of *Journal of Physics: Conference Series* have been peer reviewed through processes administered by the proceedings Editors. Reviews were conducted by expert referees to the professional and scientific standards expected of a proceedings journal published by IOP Publishing.

Journal of Physics: Conference Series **359** (2012) 012001 doi:10.1088/1742-6596/359/1/012001

A survey of the Italian research in solid state physics by infrared spectroscopy with electron-beam sources

S. Lupi[1], A. Nucara[2], A. Perucchi[3], M. Cestelli Guidi[4], E. Chiadroni[4], M. Ferrario[4], M. Ortolani[5], L. Baldassarre[3], D. Nicoletti[2], C. Mirri[4], F. M. Vitucci[2], P. Di Pietro[2], U. Schade[6], and P. Calvani[2]

[1]CNR-IOM and Dipartimento di Fisica, Università La Sapienza, Roma, Italy

[2]CNR-SPIN and Dipartimento di Fisica, Università La Sapienza, Roma, Italy

[3]Sincrotrone Trieste, Area Science Park, Basovizza, Italy

[4]Laboratori Nazionali INFN di Frascati, Frascati, Italy

[5]CNR-IFN, Via del Cineto Romano, Roma, Italy

[6] Helmholtz Zentrum Berlin GmbH., Berlin, Germany

paolo.calvani@roma1.infn.it

Abstract. Two beamlines for the use of InfraRed Synchrotron Radiation (IRSR) are presently working in Italy, at the ELETTRA storage ring in Trieste, and at the collider DAFNE in Frascati. A third facility, SPARC in Frascati, has been equipped for the extraction of Terahertz radiation with the aim to perform pump-probe time-domain experiments. Here we describe those apparata and we review the main results that the Italian groups have obtained therein, in recent years, in solid-state physics. We also describe the experiments performed in collaboration with the storage ring BESSY in Berlin, for the exploitation of Coherent Synchrotron Radiation, and the terahertz beam line recently implemented at SPARC, the Free Electron Laser of Laboratori Nazionali di Frascati

1. The beamline SINBAD at DAFNE and the high-pressure measurements

SINBAD (Synchrotron Infrared Beamline At DAFNE) was the first InfraRed Synchrotron Radiation (IRSR) beamline built in Italy, under a collaboration between the Istituto Nazionale di Fisica Nucleare (INFN) and the Università di Roma La Sapienza. SINBAD collects in a parasitic regime the radiation emitted by a bending magnet of DAFNE, an electron-positron collider that is designed to work at 0.51 GeV with a beam current of more than 1 A in each ring. Due to such a high current, the infrared extracted from DAFNE is more brilliant than that of a black body by more than one order of magnitude at 100 cm^{-1}. The front end of SINBAD collects a solid angle of 18 x 45 $mrad^2$ and the plane extraction mirror is placed at about 4,5 m from the source located along the bending magnet arc. The beamline layout, designed with the help of ray-tracing simulation, consists of six mirrors that first focus the radiation, and then transfer a collimated beam to the experimental area.

After the commissioning in 2003, two experimental stations have been built on SINBAD, both of which are equipped with Michelson interferometers. One of them includes an infrared microscope Hyperion 3000 coupled to a Focal Plane Array (FPA) detector and is devoted to chemical and biophysical studies. The other one, equipped with a cryogenic and a high-pressure apparatus, is specialized in solid-state physics. Both stations host every year numerous Italian and European groups under EU projects.

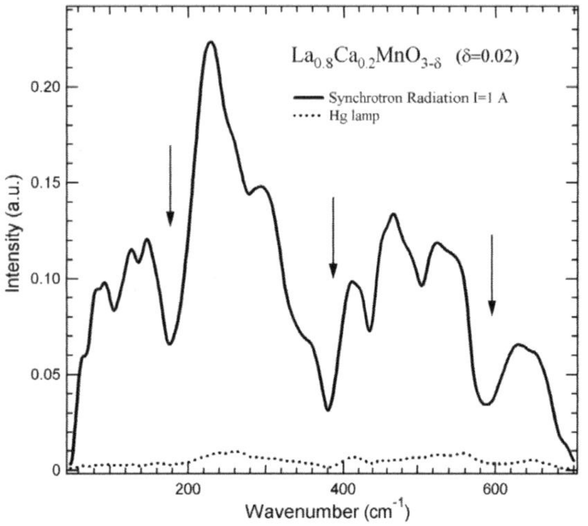

Figure 1. Comparison between the transmittance spectrum of a manganite in a diamond anvil cell (DAC) with use of the IRSR of SINBAD (solid line) and the internal Hg source of the interferometer (dashed line). The two intensities are plotted on the same scale. The arrows indicate the phonon lines.

IRSR displays all its potential in the far-infrared range and on samples of small size (Figure 1). For this reason, since the beginning the SINBAD radiation was applied to Diamond Anvil Cells (DAC) in order to study the insulator-to-metal (IMT) transitions induced by pressure in novel materials like the colossal-magnetoresistance manganites. Figure 2 shows indeed a IMT in a La-Ca manganite which is insulating at ambient pressure. Evidence for the IMT is provided by the weakening of the phonon peaks with pressure at room temperature, due to the shielding effect of the free carriers.

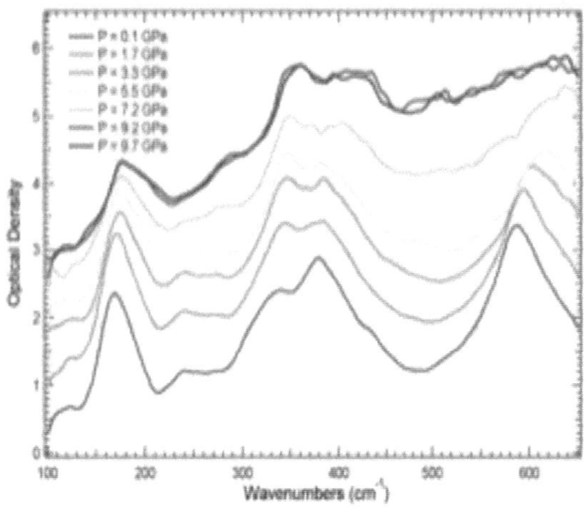

Figure 2. IMT in $La_{0.8}\ Ca_{0.20}MnO_3$, induced by pressure and observed in the FIR with IRSR [2].

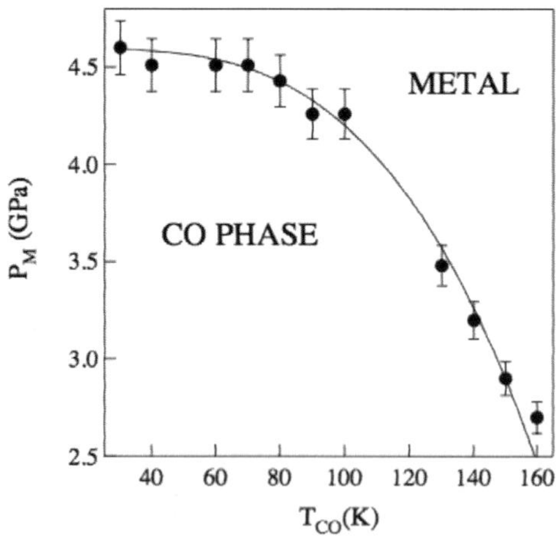

Figure 3. P-T diagram of the charge-order/metal coexistence in $Nd_{0.5}Sr_{0.5}MnO_3$ [3].

Recently, similar measurements were extended to low temperatures at SPRING-8, with the collaboration of the group of the Kobe university. Data on a Nd-Sr manganite, which is insulating for charge order (CO), allowed some of us to obtain the P-T phase diagram shown in Figure 3, which is the first to our knowledge for the CO-metal coexistence. These measurements, like those in Figure 2, could never be performed with a sufficient signal-to-noise ratio without the use of IRSR.

2. The beamline SISSI at ELETTRA and the spectroscopy of strongly correlated systems

The SISSI beamline at ELETTRA was designed and built by the group of La Sapienza, with the technical and financial support of Sincrotrone Trieste, and is operating since 2005. Its layout is much simpler than that of SINBAD and is shown in Figure 4. The extraction mirror M1 collects the beam, 65x25 mrad wide, from the bending magnet. The radiation is then focused by the ellipsoid M2 beyond the shielding wall, reflected by the plane mirror M3, and refocused by the ellipsoid M4 on the entrance of the interferometer [4].

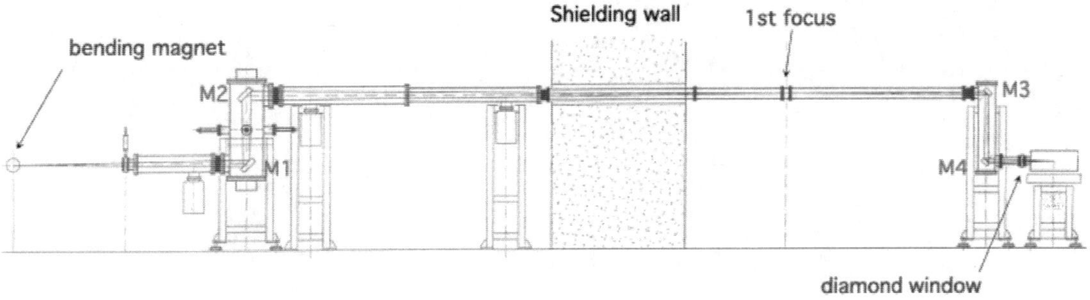

Figure 4. Optical layout of the SISSI beamline at ELETTRA [4].

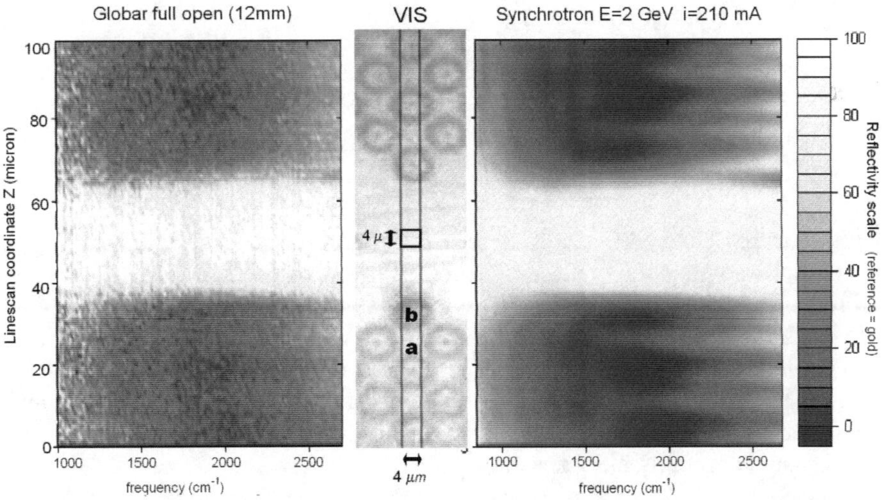

Figure 5. Mid-IR reflectance spectra acquired with a linear scan (step 1 micron) in the *z* direction on a Si_3N_4 "flower" pattern over a gold background (visible image in the center). The value of the reflectance as a function of *z* and frequency is indicated by the color scale. The knife edge aperture was 4x4 micron, the resolution was 8 cm^{-1} and 64 scans were accumulated using both the globar and the synchrotron.

The beamline feeds two experimental stations, both of which are equipped with a Michelson interferometer. As for SINBAD, one of them is devoted to chemistry and biophysics and includes an infrared microscope Hyperion 3000 with FPA detector. The other one, equipped with cryogenics and high-pressure cells, is devoted mainly to solid-state physics.

The advantage of using synchrotron radiation instead of conventional sources, when operating an infrared microscope in the mid infrared, can be appreciated in Figure 5. Therein, the spectra of a sample obtained by depositing a low-reflectance Si_3N_4 "flower" pattern on a gold substrate (shown in the middle) are measured with the microscope coupled to a MCT detector, by using either the globar (left) or the IRSR of SISSI (right). Not only the quality of the right image is much higher, but therein only one can resolve (at 2500 cm^{-1} and above) the reflectivity gap between the "flowers".

The activity at SISSI in solid state physics has been focused mainly on the study of superconductors, of strongly correlated systems, and of the transport properties at interfaces.

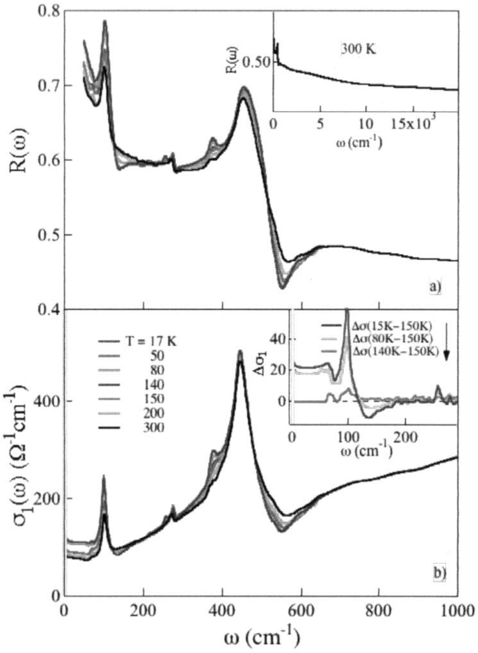

Figure 6. Far-infrared reflectivity (a) and optical conductivity (b) of undoped, polycrystalline, SmOFeAs. The inset of a) shows the reflectivity at 300 K between 50 and 20000 cm^{-1} , that of b) the difference $\Delta\sigma_1=\sigma_1(\omega,T)-\sigma_1(\omega,150\ K)$ [5].

The former subject has been addressed, in recent years, by measuring the optical properties of the newly discovered superconductors of the Fe-As family. A complete Infrared-Raman study of the phonon spectrum has been performed on SmFeAsO polycrystalline samples, both doped and undoped, and compared with Density Functional Theory calculations within the pseudopotential approximation. In this way, the structural and dynamical lattice properties of both the SmFeAsO and the prototype LaFeAsO compounds have been obtained [5]. The good agreement between the data and the optical phonon frequencies computed at the Γ point showed the capability of the employed *ab-initio* methods to describe the dynamical properties of these novel superconductors.

Extended studies on strongly correlated systems have also been performed at SISSI, with focus on the Magnéli phases of vanadium oxide: VO_2, V_2O_3, and V_3O_5. A first study was carried out by combined Raman, IR transmittance and IR reflectivity measurements at room temperature, on monoclinic VO_2 over the 0–19 GPa and 0–14 GPa pressure ranges, respectively (figure 7). Both the lattice dynamics and the optical gap showed a remarkable stability up to $P^* \approx 10$ GPa, whereas subtle modifications in the V ion arrangements within the monoclinic lattice, together with the onset of a metallization process via band gap filling, were observed for $P>P^*$. While at $P=0$ the VO_2 metallic phase is found only in conjunction with the rutile structure above 340 K, the study showed that a new metallic phase within a monoclinic structure is accessible in the high pressure regime at room temperature.

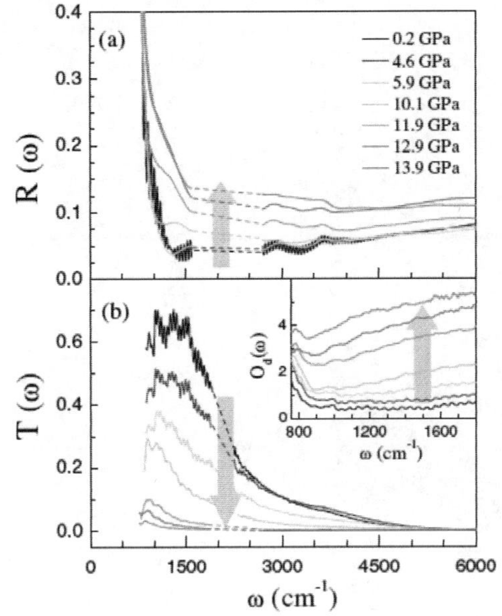

Figure 7. Mid-infrared reflectivity (a) and transmittance (b) of VO_2 at selected pressures. The. arrows indicate increasing pressure. Dashed lines replace the data at the frequencies where the diamond-window absorption of the DAC is too high. Inset: corresponding low-frequency optical density [6].

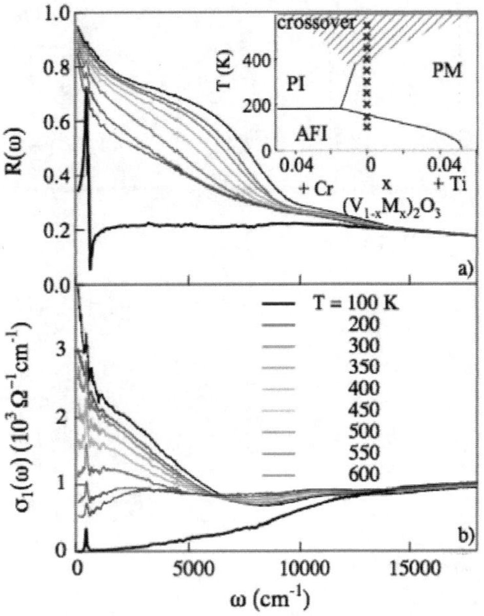

Figure 8. Metal-Insulator transition induced by decreasing temperature in V_2O_3. The inset shows its phase diagram, where AFI is the antiferromagnetic insulator, PM the paramagnetic metal, and PI the paramagnetic insulator, with the temperatures (crosses) probed by the infrared spectra shown in b) [7].

The infrared study of V_2O_3 [7] was performed both on pure and on Cr-doped crystals. In this way one could monitor the transition from the antiferromagnetic (AF) to the paramagnetic (PM) state either when it is coupled to insulator-to-metal transition (MIT), as in the former case, or not (as Cr-doped samples). The opening of a large insulating gap at 100 K is clearly shown in Figure 8, together with an upturn in the low-frequency conductivity above 400 K which indicates the appearance of a pseudogap. This feature is probably related to the destruction of quasi-particles predicted by theory when, in correlated metals, the renormalized Fermi energy becomes comparable with the thermal energy.

In the same field, a recent work on $(V_{0.989}Cr_{0.011})_2O_3$ [8] can illustrate the synergetic opportunities offered by the simultaneous presence of IR, UV, and X-ray beamlines in a facility like ELETTRA. $(V_{0.989}Cr_{0.011})_2O_3$ provides the opportunity to span all phases of the vanadium oxide through two MITs. One, from the AF insulator to the PM metal, also shows a change of crystal symmetry, the other (from PM metal to PM insulator) is iso-structural and a prototype of the correlation-driven Mott transition. By combining three spectroscopies with different lateral resolution, IR, scanning photoemission microscopy (SPEM) and X-ray diffraction (XRD), with LDA + DMFT calculations, the authors could explore the phase diagram of Cr-doped V_2O_3 from macroscopic to microscopic scales, finding evidence of an electronic phase separation in the PM phase (Figure 9). The poor metallic behavior, which partially prevents the application of this material, is therefore explained in terms of a coexistence of metallic and insulating domains stabilized by structural defects. However, both IR and XRD measurements demonstrate that external pressure can drive the MIT much closer to a genuine Mott transition, leading eventually to the homogeneous metallic phase which is required for the applications.

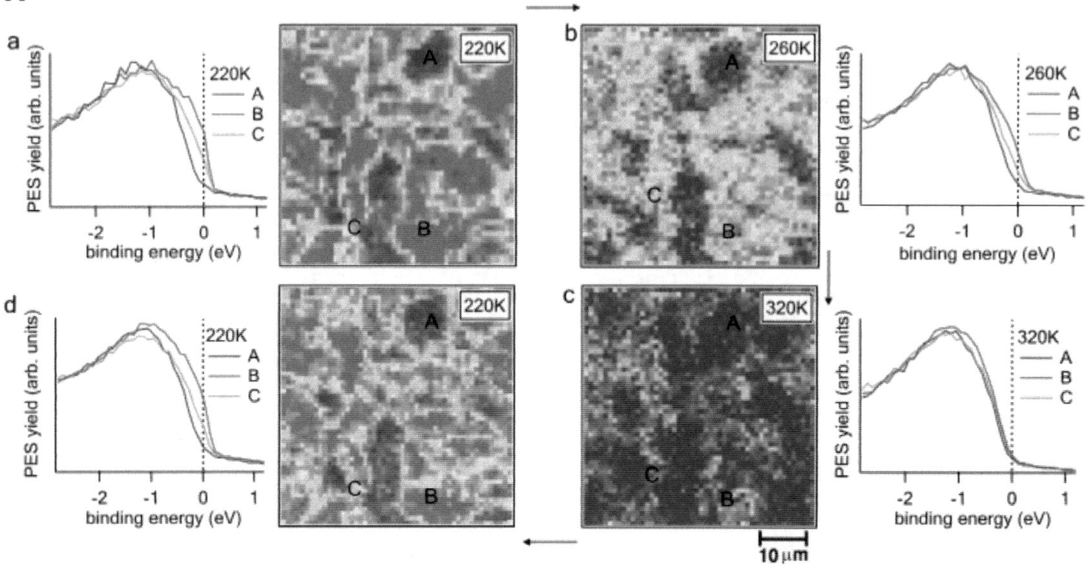

Figure 9. Scanning photoemission microscopy images and spectra on $(V_{0.989}Cr_{0.011})_2O_3$, collected at 27 eV photon energy and at different temperatures on a 50 x 50 μm sample area. The pictorial contrast between metallic and insulating zones is obtained from the photoemission intensity at the Fermi level from its minimum value, in violet, to its maximum, in red. Inhomogeneous properties are found within the PM phase at T = 220 K (**a**) and 260 K (**b**), where metallic (in red) and insulating (in blue) domains coexist. After heating the sample to 320 K (**c**) a homogenous insulating state is obtained. After a whole thermal cycle (**d**) the metallic regions can be found in the same position and shape as in (**a**).

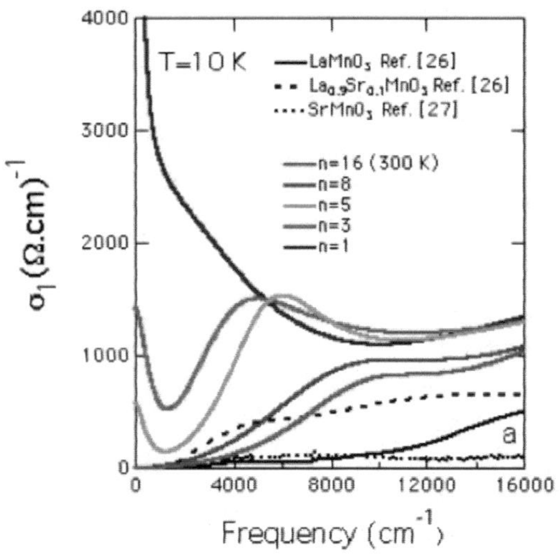

Figure 10. The optical conductivity at 10 K of the $(SrMnO_3)_n/(LaMnO_3)_{2n}$ heterostructures, showing a metallic behavior (Drude term) for $n \leq 3$.

A third solid-state subject investigated at SISSI concerns the electronic properties of the interfaces and their differences from those of bulk materials, with focus on the manganites. It is known from transport measurements that thin films made of $(SrMnO_3)_n/(LaMnO_3)_{2n}$ superlattices (SL), for small n, become metallic even if the individual layers should be insulating. At SISSI the reflectivity of such SL was measured for n=1, 3, 5, and 8 and for 10 < T < 400 K [9]. The spectra showed a T-dependent IMT for $n \leq 3$, driven by the softening of a polaronic mid-infrared (MIR) band which, in turn, built up a strong Drude term. At $n = 5$ that softening is incomplete, while at the largest SL period ($n = 8$) the MIR band was independent of T and the SL remained insulating. One could thus first observe the IMT in a manganite in the absence of the disorder, due to chemical doping, which affects the bulk transitions. Moreover, in no way the SL optical properties could be reconstructed from those of the original bulk materials. This suggests that the $(SrMnO_3)_n/(LaMnO_3)_{2n}$ heterostructures represent an original physical state of matter.

3. Sub-Terahertz coherent radiation at BESSY and the superconducting gap

A fruitful collaboration was established in 2005 between the group of Rome La Sapienza and U. Schade, who is responsible for the infrared beamline IRIS at BESSY (Berlin). This storage ring has the unique capability to work, routinely for users twice a year, in the so-called α-mode [10]. In such a condition of the magnetic lattice, very short bunches carrying a total current of about 20 mA run along an ideal orbit. Under these conditions, coherent radiation is emitted at wavelengths larger than the bunch length (figure 11), namely, between 3-5 and 30 cm^{-1} (sub-Terahertz range). The collected radiation is more brilliant than the ordinary IRSR by orders of magnitude (figure 12) and is extraordinarily stable.

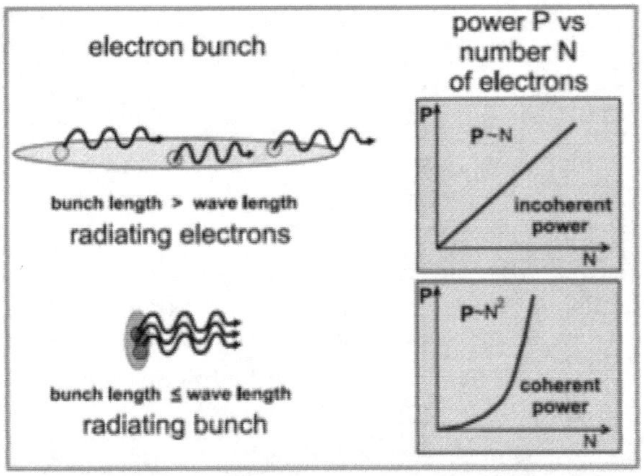

Figure 11. The principle of CSR: at wavelengths smaller than the bunch length, all the N electrons in the bunch emit coherently, and the sub-Terahertz power becomes proportional to N^2.

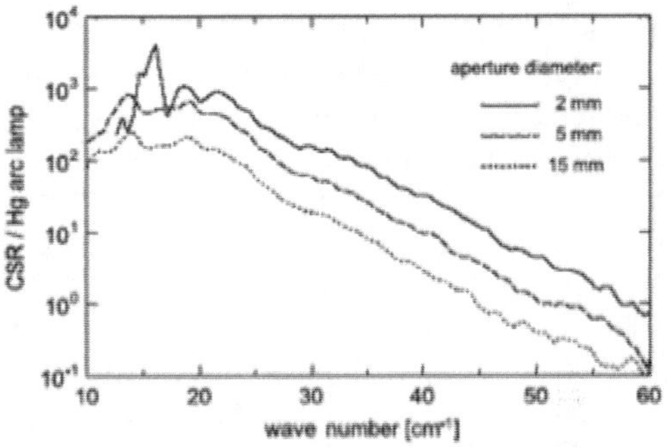

Figure 12. The effective gain of CSR with respect to a mercury lamp measured in the sample compartment of the interferometer.

Those properties of CSR were exploited in a series of experiments devoted to study the superconducting optical gap 2Δ of novel superconductors [11-13], especially those where the low critical temperature, and the consequent low value of 2Δ, prevented the use of conventional sources (including IRSR). Other phenomena related to superconductivity were also investigated, like the Josephson Plasma Resonance (JPR).

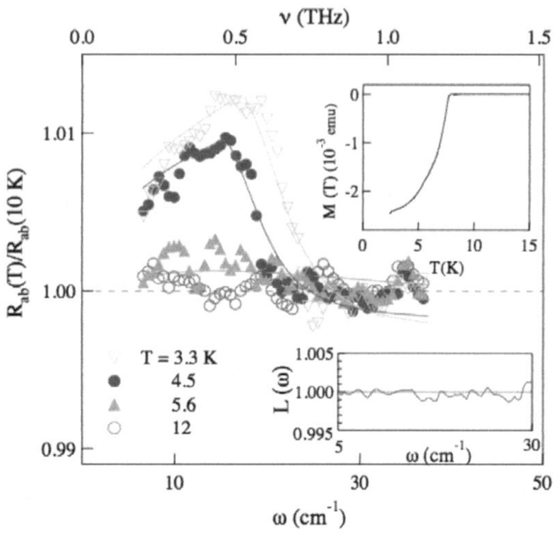

Figure 13. A 1% variation in the reflectivity ratio of CaAlSi between the superconducting and the normal phase allows one to measure the superconducting gap in this material by CSR radiation. The inset shows the stability of CSR within a few parts in 10^{-3} through the ratio of subsequent spectra (30 scans each) [11].

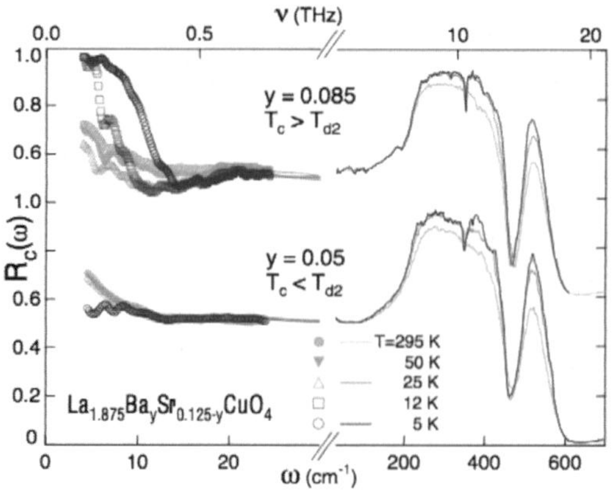

Figure 14. The Josephson plasma resonance observed by CSR radiation polarized along the c axis of the superconducting crystal La-Ba-Sr-Cu-O [12].

A first example, shown in Figure 13, is the determination of 2Δ in the superconductor CaAlSi, whose structure is similar to that of MgB_2, from the peak in the ratio between the reflectivity measured below T_c and that measured in the normal phase [12]. As shown in the inset, the overall variation of 1% can be detected thanks to the stability of CSR, which reaches a few parts per thousand between subsequent spectra. Figure 13 shows instead the JPR revealed by the CSR polarized along the *c* axis of a superconducting anisotropic crystal, the cuprate La-Ba-Sr-Cu-O. JPR is a direct insulator-to-superconductor transition which is triggered in the *c* axis by the superconducting transition in the orthogonal ab planes. In fact, the Cooper pairs of the Cu-O planes tunnel along the *c* axis through the insulating La-Sr(Ba) planes. As one can see in figure 14, while the far-infrared reflectivity remains that typical of an insulator at all temperatures, as shown by the unshielded phonon peaks, the reflectivity in the sub-Terahertz region jumps to 1 below Tc [13].

4. The Terahertz radiation source at the SPARC facility.

All the work presented up to now has concerned frequency-domain spectroscopy, even if mainly based on the use of unconventional infrared sources. Presently, the opportunity to perform time-domain spectroscopy in the Terahertz spectral region, and in particular THz-pump/THz-probe experiments, is offered to the Italian spectroscopic community. Indeed, a dedicated beamline for coherent THz radiation has been built at SPARC [14] on the "dogleg" line (see Fig. 15) and is under characterization under the project TERASPARC.

The linac-driven THz radiation is produced at SPARC as pulsed Coherent Transition Radiation (CTR) generated by the highly relativistic electron beam (100 – 180 MeV, 10 Hz repetition rate) going through the interface between the vacuum and a silicon aluminated screen placed in the vacuum pipe at 45 deg with respect to the beam direction. The backward CTR radiation, reflected normally to the beam direction, is extracted from the vacuum pipe through a z-cut quartz window and collected by a

90° off-axis parabolic mirror, whose focal plane corresponds to the source plane. The parallel beam is then reflected vertically down towards an aluminium flat mirror, placed at 45° with respect to the horizontal plane. The radiation, reflected horizontally, finally enters the home-made Martin-Puplett interferometer [15].

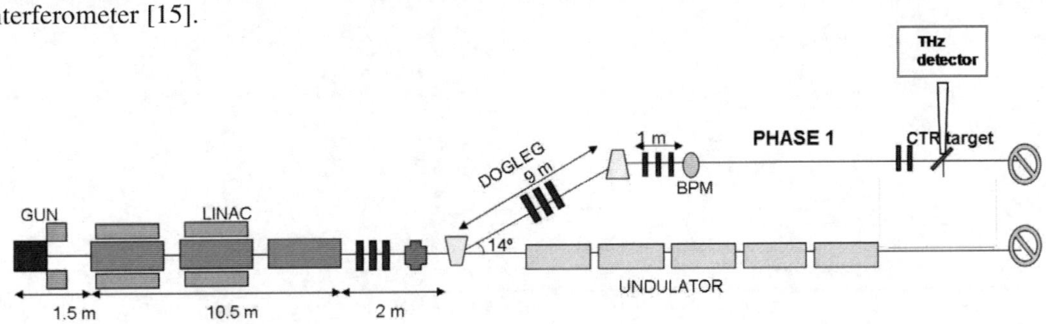

Figure 15. Layout of the SPARC facility in Frascati, with the CTR target for the extraction of Terahertz radiation.

Figure 16. The SPARC hall seen from the end of the undulators, with the THz branch on the right.

Either sub-ps high-brightness electron beams or longitudinally modulated beams can be used to generate high power coherent THz radiation, broadband or narrowband, respectively. The technique used at SPARC to manipulate such electron beam relies on low energy RF compression (the velocity bunching) [16] and on the use of properly shaped trains of UV laser pulses hitting the photo-cathode (comb laser beam) [17].

One of the interferograms measured starting from April 2011 is shown in figure 17. A 115 MeV, 200 pC longitudinally compressed beam, down to 720 (40) fs, has been measured.

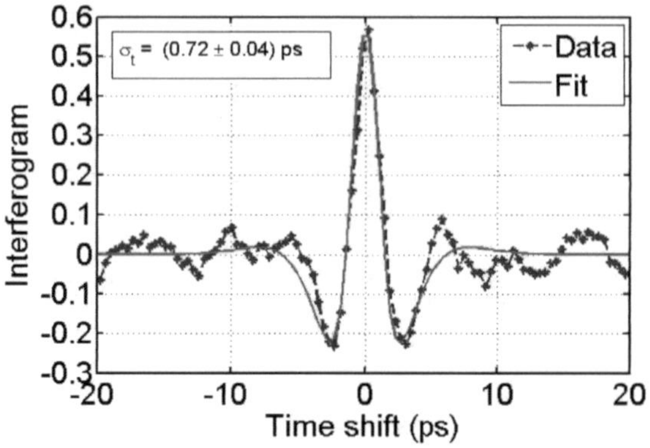

Figure 17. One of the interferograms (blue dots) collected at SPARC by means of a Martin-Puplett interferometer. The interferogram has been fit with a three-Gaussian curve (red curve) [18] to take into account the high-pass filter behavior of the interferometer. A sub-ps, 200 pC, 115 MeV electron beam has been characterized.

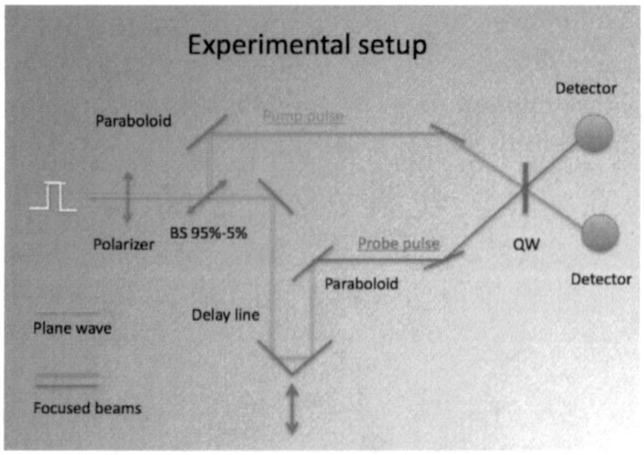

Figure 18. The experimental setup of the THz-pump/THz-probe experiments at SPARC.

The first experiment which will be performed with the new apparatus, coupled to the optical layout shown in figure 18, is the determination of the life time of the excited states in Ge-Si:Ge quantum wells. Both the pump and the probe will be in the THz range, thus reducing to a minimum the perturbation on the system. These measurements will be part of a research plan which is aimed at implementing a new class of Terahertz Quantum Cascade Lasers [19].

References

[1] Cestelli Guidi M, Piccinini M, Marcelli A, Nucara A, Calvani P and Burattini E 2005 *J. Opt. Soc. Am.* A **22** 2810.

[2] Sacchetti A, Cestelli Guidi M, Arcangeletti E, Calvani P, Marcelli A, Nucara A, Piccinini M, and Postorino P 2006 *Phys. Rev. Lett.* **96** 035503.

[3] Maselli P, Nicoletti D, Nucara A, Irizawa A, Shojij K, Nanba T, and Calvani P 2010 *EPL* **92** 47001.

[4] Lupi S, Nucara A, Perucchi A, Calvani P, Ortolani M, Quaroni L, and Kiskinova M 2007 *J. Opt. Soc. Am.* B **24** 959.

[5] Marini C, Mirri C, Profeta G, Lupi S, Di Castro D, Sopracase R, Postorino P, Calvani P, Perucchi A, Massidda S, Tropeano G M, Putti M, Martinelli A, Palenzona A, and Dore P 2008 *EPL* **84** 67013.

[6] Arcangeletti E, Baldassarre L, Di Castro D, Lupi S, Malavasi L, Marini C, Perucchi A, and Postorino P 2007 *Phys. Rev. Lett.* **98** 196406.

[7] Baldassarre L, Perucchi A, Nicoletti D, Toschi A Sangiovanni G, Held K, Capone M, Ortolani M, Malavasi L, Marsi M, Metcalf P, Postorino P, and Lupi S 2008 *Phys. Rev.* B **77** 113107.

[8] Perucchi A, Baldassarre L, Nucara A, Calvani P, Adamo C, Schlomm D, Orgiani P, Maritato L, and Lupi S 2010 *Nano Lett.* **10** 4819.

[9] Lupi S, Baldassarre L, Mansart B, Perucchi A, Barinov A, Dudin P, Papalazarou E, Rodolakis F, Rueff J-P, Itié J-P, Ravy S, Nicoletti D, Postorino P, Hansmann P, Parragh N, Toschi A, Saha-Dasgupta T, Andersen O K, Sangiovanni G, Held K, and Marsi M 2010 *Nature Commun.* 1109

[10] Abo-Bakr M, Feikes J, Holldack K, Kuske P, Peatman W B, Schade U, Wüstefeld G, and Hübers H.-W. 2003 *Phys. Rev. Lett.* **90** 094801.

[11] Ortolani M, Lupi S, Baldassarre L, Schade U, Calvani P, Takano Y, Nagao M, Takenouchi T, and Karawada H 2006 *Phy. Rev. Lett.* **97** 097002.
[12] Lupi S, Baldassarre L, Ortolani M, Mirri C, Schade U, Sopracase R, Tamegai T, Fittipaldi R, Vecchione A, and Calvani P 2008 *Phys. Rev.* B **77** 054510.
[13] Ortolani M, Calvani P, Lupi S, Schade U, Perla A, Fujita M, and Yamada K 2006 *Phys. Rev.* B **73** 184508.
[14] SPARC Collaboration 2004 *Technical Design Report for the SPARC Advanced Photo-Injector*.
[15] Martin D H and Puplett E 1970 *Infrared Phys.* **10** 105.
[16] Ferrario M et al. 2010 *Phys. Rev. Lett.* **104** 054801.
[17] Boscolo M et al. 2007 *Nucl. Instrum. and Meth. A* **577** 409.
[18] Rosenzweig J et al. 1999 *Proc. 8th Advanced Accelerator Concepts Workshop*
[19] Busby Y, De Seta M, Capellini G, Evangelisti F, Ortolani M, Virgilio M, Grosso G, Pizzi G, Calvani P, Lupi S, Nardone M, Nicotra G and Spinella C 2010 *Phys. Rev.* B **82** 205317.

Toward highly stable Terahertz Coherent Synchrotron Radiation at the synchrotron SOLEIL

J. Barros, L. Manceron, J.-B. Brubach, G. Creff, C. Evain, M.-E. Couprie, A. Loulergue, L. Nadolski, M.-A. Tordeux and P. Roy

Synchrotron SOLEIL, l'Orme des Merisiers, Saint-Aubin, BP 48, 91192 Gif-sur-Yvette, France

pascale.roy@synchrotron-soleil.fr

Abstract. The production of steady-state Coherent Synchrotron Radiation has been recently achieved in SOLEIL synchrotron using low-alpha configurations. Although the increase in THz intensity is advantageous for FTIR measurements, the signal-to-noise ratio is deteriorated by intensity fluctuations and aperiodic noise. However, we demonstrate here the possibility to use CSR as a source for FTIR spectroscopy at high resolution (10^{-3} cm^{-1}) by using a noise correction system based on double detection. Combined to the high stability optics of the infrared beamline AILES, it allows to extend the range of optimal use of the synchrotron source down to ~0.2 THz (7 cm^{-1}).

1. Introduction and background

The infrared radiation produced in a synchrotron is a powerful source that combines the advantages of a high brilliance and a wide spectral coverage. Aiming at enhancing these performances in the THz range, a number of synchrotron radiation facilities have studied Coherent Synchrotron Radiation (CSR). The longitudinal coherence is obtained either by imprinting a laser-induced modulation in the electron bunches [1-2] or by reducing the bunch length so that it becomes comparable to the emitted wavelength [3-4]. The latter technique is used in SOLEIL: the bunch length is varied by changing the value of the momentum compaction factor α.

Although this operation mode allows a huge flux enhancement in the THz range, the interaction of the electrons with their own electromagnetic field (wakefield) causes a microstructure in the electron bunch, in turn resulting in bursts of THz radiation [5]. The observed structures in the temporal THz signal lead to white noise and artifacts in the spectra after a Fourier Transform. The reported applications of CSR mostly exploit the intensity increase of this source for low resolution studies, where the stability has less impact on the signal-to-noise ratio. Meanwhile, for high resolution measurements, the use of CSR is greatly hampered by these microbunching instabilities [6].

In order to obtain an adapted CSR signal for high resolution measurements, we first determined the best operating conditions of the storage ring (described in section 2), and then developed a system to correct the remaining noise that we will introduce in part 3 of this paper. We tested this system by recording the spectrum of N_2O with three different conditions: with incoherent synchrotron radiation (ISR), with CSR, and finally with CSR and the noise correction system. Results are given in part 4.

Journal of Physics: Conference Series **359** (2012) 012002 doi:10.1088/1742-6596/359/1/012002

2. Production of Coherent Synchrotron Radiation

2.1. Optimum operating parameters

At SOLEIL the typical bunch length is of 13.8 ps for a 3 MV radiofrequency cavity voltage, and corresponds to a momentum compaction factor of $\alpha_0 = 4.34 \cdot 10^{-4}$. It could be shortened down to 1 ps using low alpha configurations.

The dependence of the signal-to-noise ratio on the current and the bunch length was first studied with one stored bunch, on 11 values of α ranging from α_0 to $\alpha_0/100$. We chose a value of $\alpha_0/25$ where the CSR was observed from 7 to 21 cm^{-1} and remained stable in a sufficient current range (0.055 to 0.105 mA) to allow a multi-bunch operation. For 10 different values of the current in the single bunch, the signal-to-noise ratio between 7 and 21 cm^{-1} was calculated on low resolution spectra (0.5 cm^{-1} resolution) for 1 scan. While the typical signal-to-noise ratio for a normal, incoherent operation in the same conditions is 1.6, it could reach 30 for I = 0.064 mA and $\alpha_0/25$.

2.2. Intensity enhancement

The multi-bunch operation was then performed with 0.064 mA per bunch and 416 bunches. Figure 1 shows an intensity enhancement over 10^4, which was measured in the 7-21 cm^{-1} range.

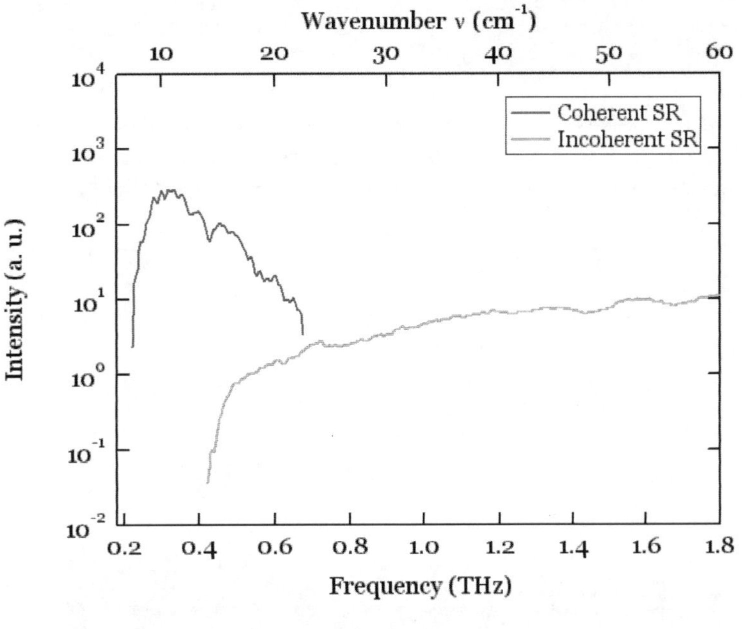

Figure 1. Gain in intensity of the CSR compared to the incoherent radiation. Green line: multi-bunch operation, $\alpha = \alpha_0$ with 400 mA total ring current. Red line: multi-bunch operation, $\alpha = \alpha_0/25$ with 26.8 mA total ring current.

Since the choice of the operating conditions is a compromise between a low level of noise and a high intensity gain, it is worth noticing that a gain of 10^5 over a wider frequency range can be reached for a total ring current over 27 mA. Although the level of noise then prevents any use at high resolution, this source is of high interest for condensed matter studies.

3. Correction of remaining noise

All the measurements were performed on the high resolution branch of the AILES beamline [7-8] of SOLEIL. Knowing that the transport optical elements and the pumping system of AILES were designed to avoid noise, the key issue is thus the reduction of the effect of the source fluctuations. Consequently, we developed a noise correction system as shown in Figure 2. The infrared spectra were acquired using a Bruker IFS125 interferometer coupled to two composite bolometers cooled at 4.2 K. A 6 μm Mylar beamsplitter was used to split the signal into a reference beam (less than 10% of the total intensity) and a transmitted beam, detected simultaneously.

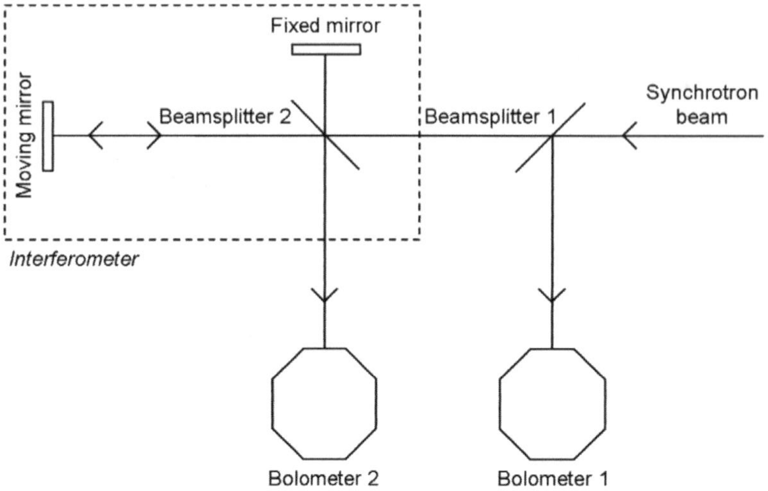

Figure 2. The noise correction system. Beamsplitter 1 divides the radiation into a reference beam detected by bolometer 1, and a transmitted beam modulated by the interferometer before being detected by bolometer 2.

As the bolometers only detect the varying part of the input, the only modulation observed by the reference bolometer arises from the source noise. By subtracting this signal from the interferogram of bolometer 2, the noise components are corrected. In order to make sure that the two detectors match, two identical optical filters, as well as identical electronic high-pass filters have been installed. Figure 3 is an example of a corrected CSR signal; the gains in signal-to-noise ratio typically range from 1.2 to more than 10.

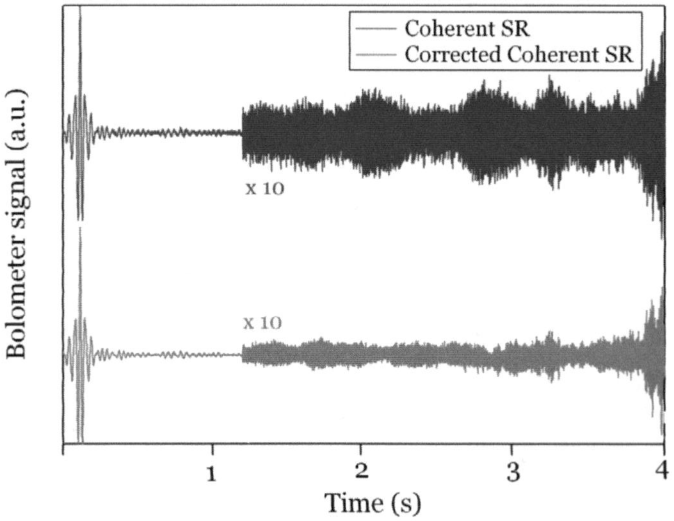

Figure 3. Noise correction on an interferogram obtained with CSR, 416 bunches, $\alpha = \alpha_0/25$ and 26.6 mA total ring current.

4. Tests at optimum resolution

The feasibility of high resolution measurements with corrected CSR was determined using nitrous oxide N_2O. This molecule has regular pure rotational lines in the frequency range of interest [9-11]. For our experiments, a single pass White-type cell set on a 0.86 m optical path was filled with 3.5 mbar of N_2O. In these conditions we barely expected the absorption bands to arise from the noise, since the $p(N_2O)xL$ product reported in previous papers recording the N_2O spectrum in the 1-100 cm^{-1} range always exceeded ours [10-11].

A test was first carried out with incoherent synchrotron radiation, $\alpha = \alpha_0$, and 400 mA that were distributed among 416 bunches. After 1h measurement at 10^{-3} cm^{-1} resolution, all the rotational lines from R(25) at 21.7760 cm^{-1} to R(49) at 41.8136 cm^{-1} could be assigned. However, the signal-to-noise ratio calculated on two successive raw spectra was under 1 in the 7-21 cm^{-1} range.

We then installed the double detection system and repeated the same measurement with CSR, and the chosen machine conditions: $\alpha = \alpha_0/25$, with 26.6 mA total ring current, distributed among 416 bunches. The effectiveness of the noise correction system permits to directly draw a comparison between ISR and the corrected CSR. A gain of $6{\cdot}10^4$ in signal-to-noise ratio was calculated, as presented in Figure 4. Also shown is the comparison between corrected and uncorrected CSR; for a given frequency, the signal-to-noise ratio of the corrected CSR could reach 8 times the uncorrected value. The noise reduction allowed assigning two additional transitions at the limits of the considered frequency range.

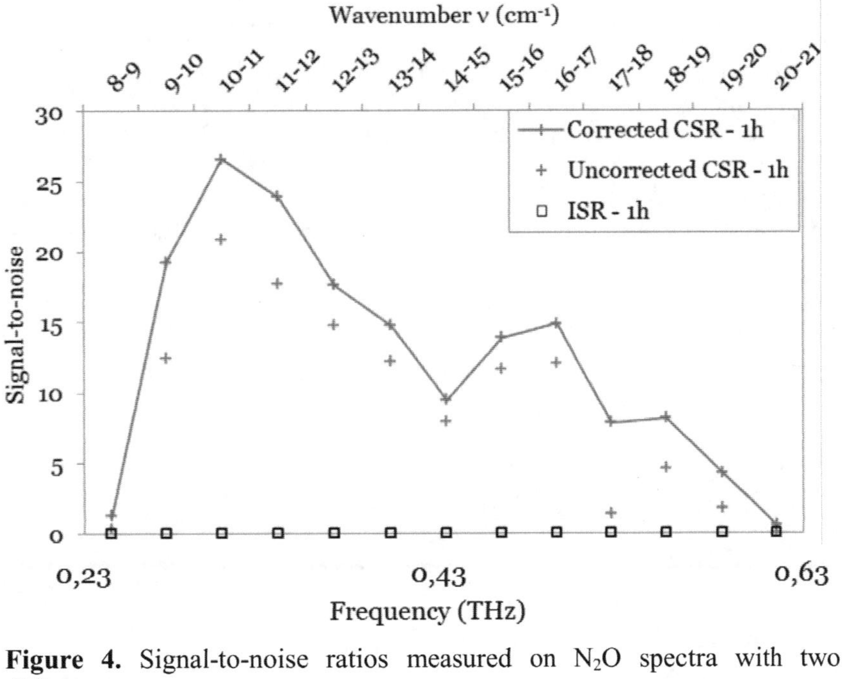

Figure 4. Signal-to-noise ratios measured on N_2O spectra with two different sources.

After 1h of measurement using the corrected CSR, the signal-to-noise was sufficient to assign without ambiguity the transitions between R(10) and R(23). The absorption spectra thus proved the complementarity of the two emission modes in terms of usable frequency range.

5. Conclusion and outlook

The high intensity and signal-to-noise enhancement observed on the AILES beamline using the coherent synchrotron radiation and a double detection system are a strong indication that this source can now be optimally exploited for high resolution spectroscopic applications. Combined with the performances of AILES in the mid- and far-infrared ranges, the use of the CSR will provide to the users of the beamline a stable and brilliant source from 7 to 3000 cm^{-1}. This enhanced stability may also favour other demanding applications, such as near-field THz spectromicroscopy.

References

[1] Bielawski S, Evain C, Hara T, Hosaka M, Katoh M, Kimura S, Mochihashi A, Shimada M, Szwaj C, Takahashi T and Takashima Y 2008 *Nature Physics* **4** 390
[2] Byrd J M, Hao Z, Martin M C, Robin D S, Sannibale F, Schoenlein R W, Zholents A A and Zolotorevet M S 2006 *Phys. Rev. Lett.* **96** 164801
[3] Byrd J M, Leemans W P, Loftsdottir A, Marcelis B, Martin M C, McKinney W R, Sannibale F, Scarvie T and Steier C 2002 *Phys. Rev. Lett.* **89** 224801
[4] Abo-Bakr M, Feikes J, Holldack K, Kuske P, Peatman W B, Schade U and Wüstefeld G 2003 *Phys. Rev. Lett.* **90** 094801
[5] Venturini M and Warnock R 2002 *Phys. Rev. Lett.* **89** 224802
[6] McKellar H R W 2010 *J. Mol. Spec.* **262** 1
[7] Brubach JB, Manceron L, Rouzières M, Pirali O, Balcon D, Kwabia Tchana F, Boudon V, Tudorie M, Huet T, Cuisset A and Roy P 2010 *AIP Conf. Proc.* **1214** 81
[8] Roy P, Rouzières M, Qi Z, Chubar O 2006 *Infr. Phys. Tech* **49** 139
[9] Morino I, Yamada K M T 2003 *J. Mol. Spec.* **219** 282
[10] May T E, Bergstrom J C, Dallin L O and Appadoo D R T 2008 *IRMMW-THz Conf. Proc.* 1-3
[11] Takahashi T, Matsuyama T, Kobayashi K, Fujita Y, Shibata Y, Ishi K and Ikezawa M 1998 *Rev. Sci. Instr.* **69** 3770

Journal of Physics: Conference Series **359** (2012) 012003 doi:10.1088/1742-6596/359/1/012003

IR beamline at the Swiss Light Source

Lerch Ph, Quaroni L, Wambach J, Schneider J, Armstrong D B, Rossetti D, Mueller F L, Peier P, Schlott V, Carroll L, Friedli P, Sigg H, Stutz S, Tran M†

Paul Scherrer Institute, Swiss Light Source, CH 5232 Villigen, Switzerland

† University of Geneva, Condensed Matter Physics Dept., CH 1203 Geneva, Switzerland

E-mail: philippe.lerch@psi.ch

Abstract. The infrared beamline at the Swiss light source uses dipole radiation and is designed to transport light to four experimental stations, A, B, C, D. Branch A is dedicated to far IR work in vacuum; branch B is a micro-spectrometer; branch C is dedicated to high resolution spectroscopy in the gas phase; branch D is a pump and probe set-up. This contribution describes the optical layout and provides a brief survey of currently available experimental stations. The beamline is in regular user operation since 2009.

1. Source and optical layout

The SLS storage ring (2.4 GeV, 400 mA) operates in top-up mode. Dipole radiation (39 mrad x 61 mrad) is extracted from a bending magnet ($\rho = 5.75$ m, $B = 1.4$ Tesla) using M1, a plane, slotted, water cooled Al mirror. If not in use, M1 is protected by a movable absorber. The dipole chamber was modified to accept a vacuum chamber that includes M1, as well as FM2 the first focussing element. This allows the distance between "source center" and M1 to be 820 mm. Radiation is transported along three segments (with focii F1, F2, F3) of 1:1 metal coated optics, $f/\mathrm{D} \approx 3$, at the end of which we find three equivalent focii, named ports A,B,C. From F2, a transfer line leads to port D. We perform experiments at one port at the time. Near F1, the first intermediate focal point, a wedged diamond or silicon vacuum tight window separates the UHV machine vacuum from the beamline vacuum. These windows can remotely be swapped at all times. Details about the window changer and a sketch of the optical layout are available on-line [1]. The slope error of each mirror has been specified to allow imaging in the visible part of the spectrum. This minimizes optical aberrations along the beamline. The quadratic transverse size of the light distribution at F3 (a replicate of the extended source emitted within the bending magnet) $\sigma_T^2 = \sigma_s^2 + \sigma_\phi^2 + \sigma_g^2$ includes contributions from the e-beam source, diffraction and geometrical projection, respectively [2, 3, 4]. σ_T has been estimated using the *SRW* code [5]. σ_g has usually little practical importance, whereas σ_s dominates in the NIR range [4]. IR spectroscopy using the Fourier transform technique is utmost sensitive to light source instabilities that are sampled into the interferograms and appear as spectral artifacts in the data. We performed a detailed noise study of two comparable IR ports (Soleil and SLS) [6].

2. FIR-MIR vacuum spectrometers at branch A

At branch A a mirror coupling box (see section 3 for details) feeds light to a Bruker IFS66v Fourier transform spectrometer dedicated to far IR work. Liquid nitrogen (LN) cooled MCT and Si detectors cooled to 4.2 K or 2.2 K are available. Based on a PSI prototype, the Group of Dirk van der Marel at the University of Geneva built a vacuum microscope using a pair of X15 Schwarzschild objectives. This unit fits the sample chamber of the spectrometer. At the focus of this optical system, there is a diamond anvil cell (DAC) cooled by a He_4 flow cryostat. This set-up allows IR investigations in transmission for samples kept at elevated pressure (few kbar $< P < 200$ kbar) and reduced temperature (22 K $< T <$ 300 K). Preliminary tests using SR light have demonstrated the principle of operation. Fig. 1 shows the IR spectrometer in which the system that holds the cryostat, the DAC, steering and the focusing optics is mounted. The sample is positioned with an XYZ manipulator that can be actuated under vacuum, whereas the stages holding the Schwarzschild objectives can be actuated at room pressure only. The optical system is being extended to allow investigations on reflective samples. A movable dichroic mirror is used to steer laser light and collect fluorescence light used to monitor the pressure in the DAC.

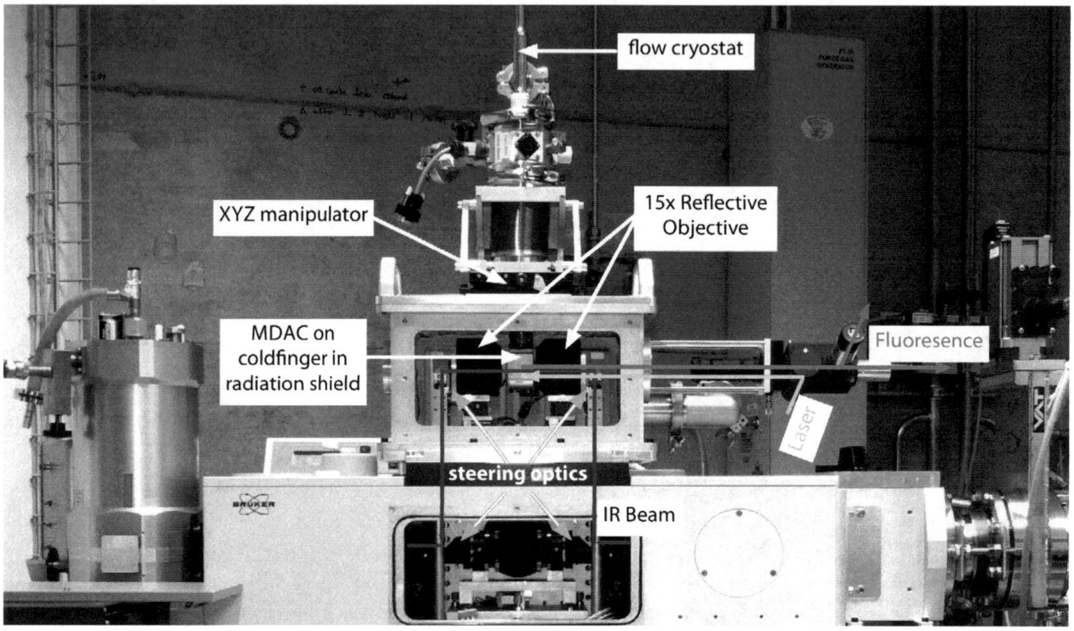

Figure 1. Vacuum micro-focus, DAC, and He_4 flow cryostat mounted on a IFS66v spectrometer.

Branch A also accepts a home-built Martin-Puplett-Interferometer (MPI), which can be operated in vacuum to avoid water absorption and has been designed to be directly flanged to the beam line output port. The sizes of the optical elements (focusing mirrors, roof mirror and wire grid beam splitters) allow beam diameters of 80 mm, which suits the MPI well for measurements in the long wavelength range (5 to 200 cm^{-1} depending on the wire grid beam splitters). The MPI control is fully integrated in the SLS EPICS control system and provides in the present configuration step-scan capability. An upgrade of the moveable roof mirror for continuous rapid scan is foreseen for the near future.

The long wavelength transmission of the IR BL port A has been determined with the MPI [7]. A liquid He_4-cooled indium antimonide (InSb) hot electron bolometer with a frequency response up to 2 THz (67 cm^{-1}) has been used as an IR detector. Figure 2 shows the interferogram (left) and the spectral distribution (right) of the incoherent synchrotron radiation. As long wavelengths ($<$ 10 cm^{-1}) are suppressed by the SLS IR BL (port A) the roll off at higher wavenumbers (> 30 cm^{-1}) can be attributed to the bolometer sensitivity and the bandwidth of the free standing wire grid beam splitters. Interferograms have been taken in air (blue curves) and at pre-vacuum pressures of $< 10^{-4}$ mbar (green curves) to show the influence of water absorption lines in this spectral range. A transfer function of the SLS IR BL (port A) has been fitted to the data measured by the MPI (green curve in fig. 3), agreeing quite well with SRW [5, 7] beam transport simulations of the beam line (red curve in fig. 3). It is worth emphasizing, that the intensity of the coherent FIR radiation extracted at port A when the storage ring is operated in the so-called "slice energy modulation" mode [8, 9] is significantly larger than in the regular operation mode. Using an electro-optical detection scheme, the "sliced" coherent FIR camshaft pulse length at port A is ≈ 850 fs FWHM (not shown).

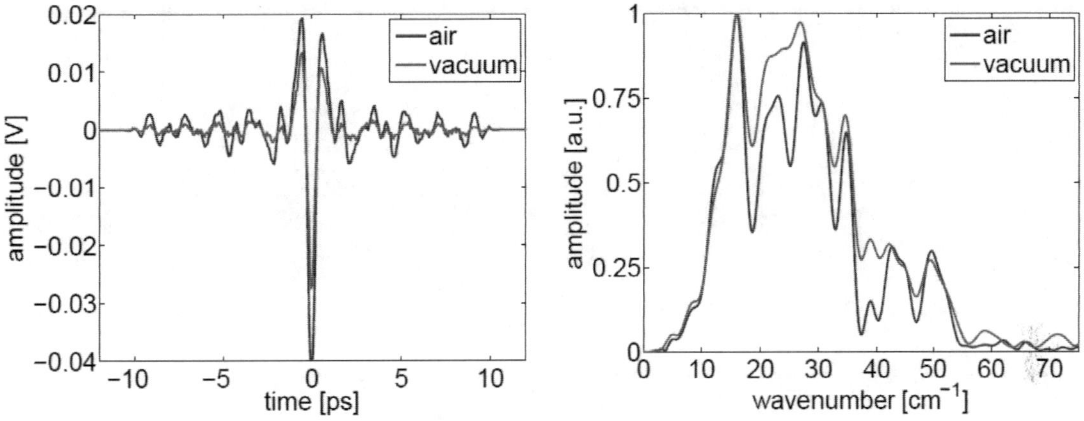

Figure 2. Left: step-scan interferogram (incoherent radiation) and right: spectrum measured with a home made Martin-Pupplet spectrometer at the SLS.

3. Mid IR spectro-microscopy at branch B

At branch B, a coupling box feeds focused light into a Bruker Vertex 70 vacuum Fourier transform spectrometer, itself coupled to a Bruker Hyperion 3000 IR microscope (m_m in table 1). The coupling box houses an elliptical mirror that reduces (m_c in table 1) the size of the light spot available at F3. From there, a $f_p = 100$ mm parabolic mirror collimates light into the interferometer unit. Modulated light is sent *directly* [10] to the Hyperion microscope equipped with the usual palette of X15, X36, X20 (Ge)ATR, and X15 GIR objectives. To estimate the actual demagnification factors m_i expected with this set-up we consider $m_i = m_c \cdot m_m$, where m_c and m_m are the contributions due to the coupling and microscope optics, respectively. The factor m_c is straightforward and depends of the focii ratio of the chosen ellipse. The factor m_m is approximated by the ratio f_p/f_m, where f_m is the actual focal length of the Schwarzschild objectives. Table 1 summarizes the situation.

The estimated and observed demagnification factors differ; $m_r > m_i$. Two comments are here in order. First, the X15 objective should actually produce a demagnification factor of 15. However, the ratio of focal distances for this optics matched to standard tube length of 160 mm gives 160 mm / 13.1 mm ≈ 12. Second, because this microscope does not use infinity corrected objectives, there is (by design) one additional intermediate focus in the light path between the interferometer and the entrance pupil of the Schwarzschild. Thus, the assumption that m_m is solely given by the focii ratio between the collimating parabolic and Schwarzschild

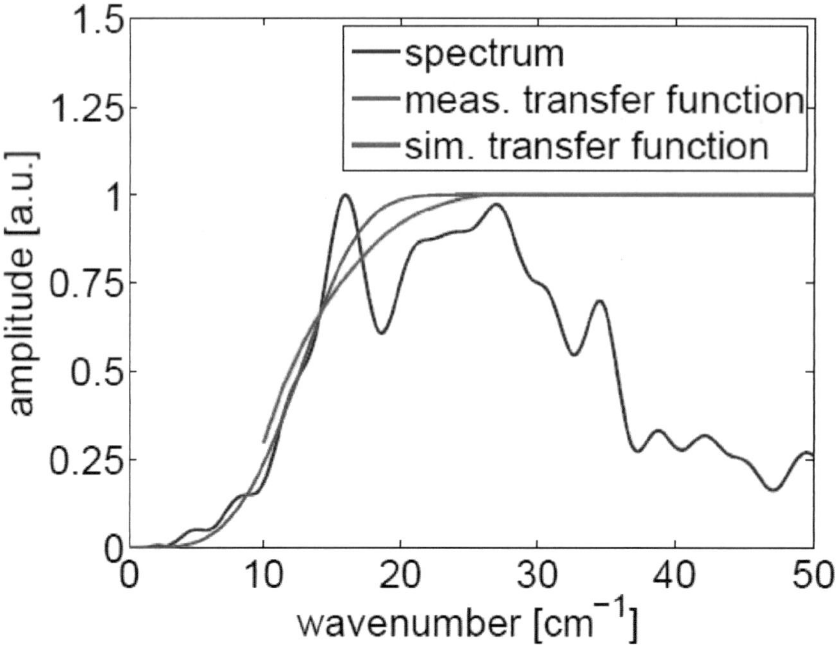

Figure 3. Measured spectrum and simulated transfer function.

Table 1. The focal length of the Schwarzschild optics $f_m = D/2NA$, NA the numerical aperture, D the size of the entrance pupil of the objective. $m_m = f_p/f_m = 100$ mm $/f_m$ assumes collimated transport to the X15 or X36 optics. The ratio of the focii of the elliptical coupling mirror is m_c. The expected demagnification factor is $m_i = m_m \cdot m_c$ whereas m_r is the observed value. Dimensions are given in mm. Differences explained in the text.

	f_m	NA	D	m_m	m_c	m_i	m_r
X 15	13.1	0.40	10.5	7.6	6	45	60
X 15	13.1	0.40	10.5	7.6	3.4	25	34
X 36	5.4	0.52	5.6	18.5	6	110	143
X 36	5.4	0.52	5.6	18.5	3.4	63	81

secondary optics is not accurate for this optical system. We determined experimentally that the demagnification factor between the spectrometer input focus and the sample focal plane to be $m_{m\,eff} = 10$ and 24, for the X15 and X36 optics, respectively. For all practical purposes, we used $m_{m\,eff} \times m_c$ to estimate m_r in Table 1. The left panel of Fig. 4 shows the wavelength dependence of the spot size, extracted from the intensity profiles (right panel) measured in transmission (X36, NA $= 0.52$, $m_c = 3.4$) by scanning a 5 μm diameter pin-hole across the beam. The slope of the experimental data is well described by FWHH $\propto 0.48 \cdot \lambda$ and agrees reasonably well with the expected diffraction limit [2, 11, 12]. The offset of the exp. data is $\approx$ 4.8 μm and includes contributions from the electron source size (visible at higher photon energy [4]) as well as the finite size of the pinhole that was used to scan the spot profiles. Strictly speaking, in order to extract precise information on the source size we should perform a 2D

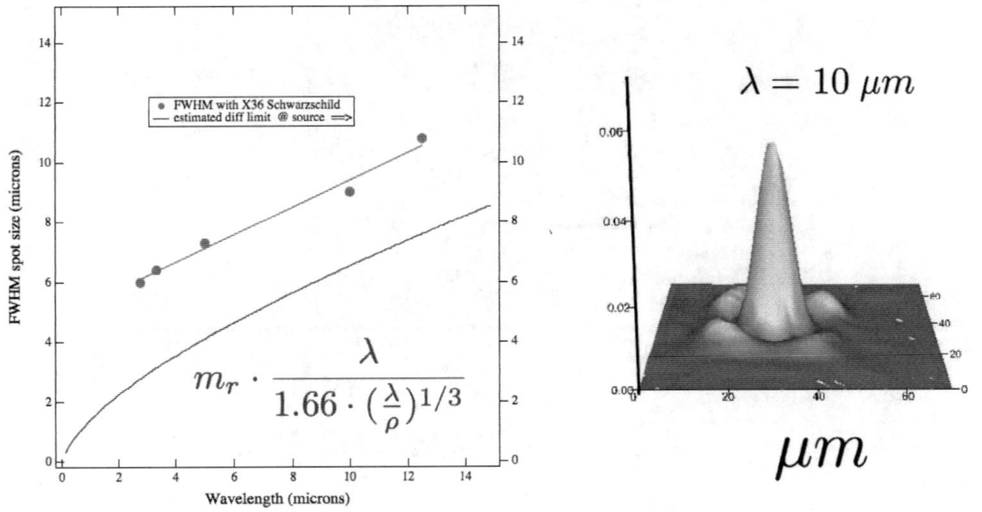

Figure 4. Left: FWHM spot size as a function of wavelength. Right: Spot intensity distribution measured (without any aperture) at 1000 cm^{-1} by scanning a 5 μm diameter pinhole across the focus of a pair of X36 Schwarzschild objectives. $m_r = 81$, $\rho = 5.75$ m.

deconvolution of the measured light profiles taking into account the point spread function of the optics and the pinhole. This shall be discussed in detail elsewhere.

This IR spectro-microsope offers a diffraction limited spot size and serves the broad community of investigators that require focused light from 400 to 4000 cm^{-1}. The instrument was further improved with a custom manufactured XYZ manipulator for positioning of the condenser Schwarzschild objective, to allow spectro-microscopy in transmission with enhanced stability. Fig. 5 illustrates this unit. The micrometric XY stage used to manipulate samples in the focal plane has been removed for clarity. The stability of the condenser unit is affected by the power dissipated by the tungsten lamps illumination system. More precisely, it is a *change* in illumination that causes long term drifts, as the system reaches a new thermal equilibrium sitation. A LED-based illuminator, or a thermal decoupling of the lamps' power supply from the optical assembly is likely to limit long term drifts. A liquid He_4 flow cryostat can be placed in the focus, for transmission and/or reflection work.

For work at elevated pressure using a diamond anvil cell (DAC), the pressure determination is often based on a shift of a fluorescence line in a crystal like ruby [13]. The excitation of the crystal and the collection of the fluorescence light is achieved "in-situ", with the DAC remaining in the IR focus. The change from "IR work" to "P-monitoring" is obtained by exploiting all the capability of the IR microscope. Our scheme does not require any modification (followed by re-alignment) of the microscope but only the placing of a tiny 2 mm diameter mirror to steers the excitation laser light along the IR path. This mirror can be positioned to avoid any IR light loss. Fluorescence light is collected through one of the ocular tubes [14].

Investigators active in cell biology, get access to a cell culture lab on-site allowing cell growth and providing Level II containment for samples of biomedical interest. A variety of enclosed sample holders allowing a controlled environment is available for spectromicroscopy experiments on solutions and contained samples, including living cells and solution processes.

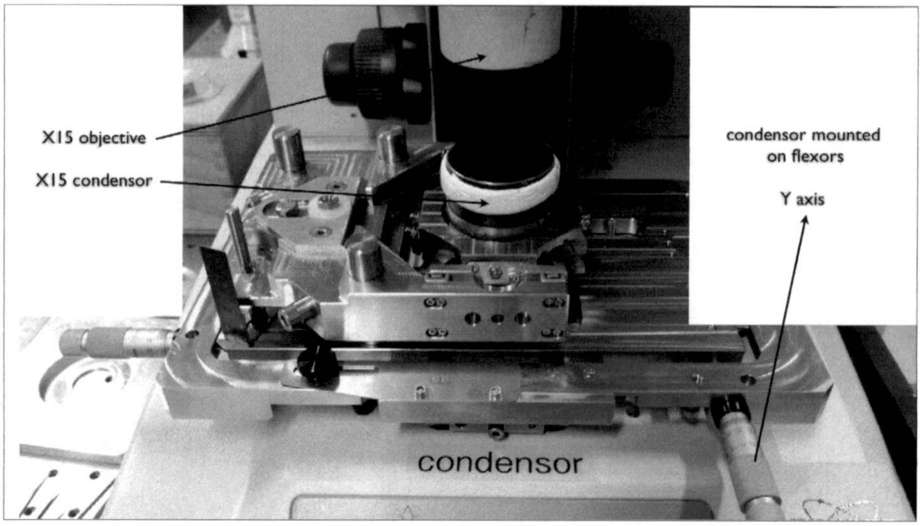

Figure 5. Custom made precision condenser manipulator. The angular position of the optics can be adjusted in two directions with flexors. The X and Y directions are adjusted by micrometric screws. The computer controlled XY sample stage has been removed for clarity.

4. Ultra high resolution at branch C

From F3C, radiation is transported through a dedicated vacuum transferline coupled to a ultra high resolution Fourier transform spectrometer Bruker HR 125. This instrument has 11 m of optical path difference and is operated by the group "Molecular Kinetics and Spectroscopy" of Martin Quack at ETH-Zurich. The transfer line between F3 and the spectrometer input was designed to re-shape the spot using cylindrical optics. Details are given in Ref [15].

5. Pump and probe at branch D

Synchrotron-based infrared pump and probe (IRPP) experiments offer a unique experimental niche in the sense that they combine focusing capability together with extended spectral and dynamic bandwidth. The available time resolution lies in the 100 ps range is well suited to study dynamics occurring in condensed matter (phase change materials, carrier diffusion, electron-hole or quasiparticle recombination) or chemical processes in which the identification of intermediate reaction species (i.e. photo catalyzes) is required.

For these experiments at branch D, light is taken from F2, and steered along a 10 m long vacuum transfer line (focussed optics) into a purged Vertex 70 Fourier transform interferometer. The coupling scheme is similar to the one used at branches A and B. The side exit of the interferometer feeds collimated light to an optical system that allows broadband pump (laser) and probe (IR synchrotron) experiments with a time resolution $\approx$ 100 ps. Proof of principle as well as a technical description of this set-up is given in Ref. [16]. We aim at developing this IRPP set-up to the level of a user instrument.

A micro-focus setup allows the investigation of samples of sizes $\approx$ 10 μm using a pair of 15X Schwarzschild optics. Simultaneous measurements in both transmission and reflection are possible. The optical path is illustrated in Fig. 6 for the standard setup (focus length of 100 mm, $\approx$ 100 μm sized beam) with parabolic mirrors instead of Schwarzschild objectives. The data acquisition system is programmed to acquire the camshaft as well as fractions of the quasi continuous part of the synchrotron emission. With this method, the response of the sample upon laser excitation gets recorded at later time stamps (> 30 ns) with a resolution given by the used detector (typically 10 ns) while maintaining the 100 ps time resolution at short delay times. Compared to a traditional time resolved step scan FTIR system using a thermal source, the (signal × bandwidth) advantage of this synchrotron based system is > 2000 when using the 100 ps long camshaft pulse. In the quasi-continuous mode, the advantage reduces to > 50. The latter number reflects the brilliance advantage of the synchrotron which remain in spite of the significant beam losses likely to take place along a complex transfer line.

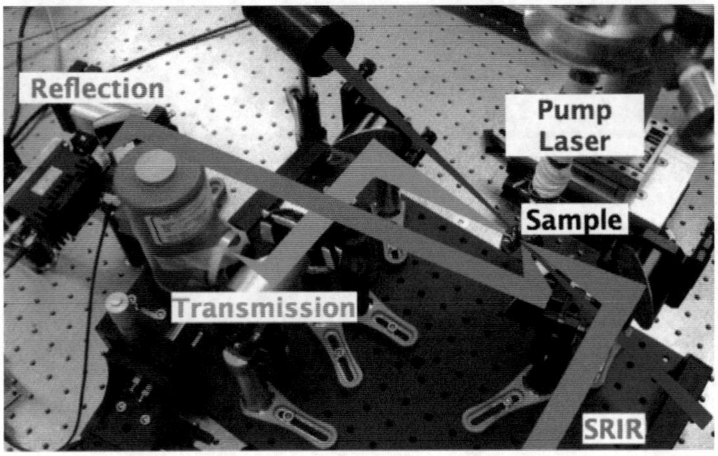

Figure 6. View of the focus region at the IRPP set-up. A similar beam path arrangement is obtained when the micro-focus unit based on a pair of Schwarzschild objectives is used.

Fig. 7 illustrates the spectral bandwidth as well as the timing capability of the entire system using the micro-focus unit. The Nd:YAG excitation pulse at 1.06 μm generates excess charge carriers in InGaAs, which lead to an excess absorption due to valence-interband and conduction-intraband transitions. We observe optical gain at energies above the InGaAs bandgap energy at 0.72 eV (see unpumped spectrum). The time evolution of this electron-hole population is probed dynamically by varying the time delay between the pump and the probe. Control of the time delay is achieved with a phase-lock loop system that synchronizes a master oscillator running on the pump laser to the SLS clock. Details are given in Ref. [16]

6. Summary

In summary, we described a versatile facility that allows at branch A to perform FIR spectroscopy in vacuum at moderate resolution and offers an easy to use experimental platform. In one such case, the high-pressure and low temperature set-up developed by the Geneva team is attached to the existing interferometer. Branch C allows FIR as well as MIR spectroscopy in vacuum at very high resolution. Branch B is the most popular platform, since it offers, in addition to a diffraction limited spot, the possibility to couple a large variety of experiments like XY sample stage, (heated) high pressure cells, micro-fluidic cells, a slim He4 flow cryostat, to name just a few. Branch D is dedicated to pump and probe experiments in the MIR range, and is still in development. A micro focus unit using a pair of X15 Schwarzschild optics (46 mm space along the optical axis) is available. Currently, we operate with synchrotron light one branch at the time.

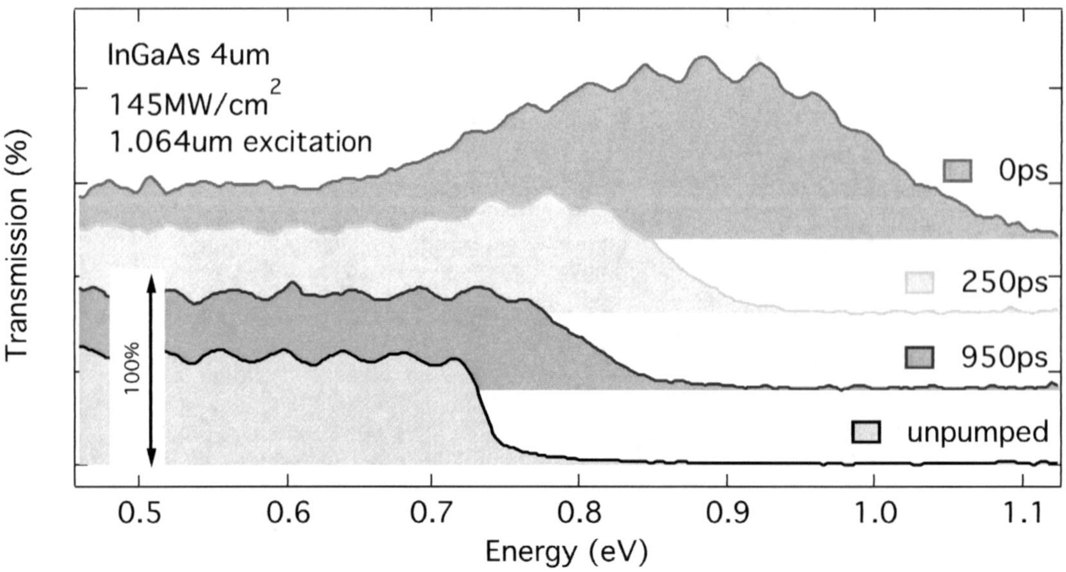

Figure 7. Normalized MIR/NIR Transmission of a 4 μm thick InGaAs layer on InP upon optical pumping with a 1.06 μm laser excitation at $\approx$ 145 MW/cm^{-2} for different pump/probe delays. The microfocus unit allows spotsizes of 16 μm (probe) and 35 μm (pump).

References

[1] Details on window changer and optical layout available: http://www.psi.ch/sls/ir/optics
[2] Hirschmugl C J, PhD Thesis, Yale University (1994)
[3] Nucara A, Calvani P, Marcelli A, Sanchez del Rio M, Rev. Sci. Instrum. **66**, 1934 (1994)
[4] Levenson E, Lerch Ph, and Martin M C, J. Synchrotron Rad. **15**, 323 (2008)
[5] Chubar O V, Elleaume P, Synchrotron Radiation Workshop code, SRW, ESRF (1999)
[6] Lerch Ph, Dumas P *et al*, to be published in J. Synchr. Rad. (2011)
[7] Müller F L, *Electro-optical bunch length measurements at the Swiss Light Source*, PhD thesis, University of Bern (2011)
[8] Khan S, Holldack K, Kachel T, Mitzner R, Quast T, Phys. Rev. Lett. **97**, 074801 (2006)
[9] Ingold G, Beaud P Johnson S L, Schlott V, Schmits T, Streun A, *A sub-ps tunable X-ray undulator source for laser/Xray pump&probe experiments at the SLS*, Technical Report, PSI (2007)
[10] The original design of the Hyperion 3000 instrument includes a coupling optics that actually *increases* the spot size in the focal plane. In our case, this coupling uses plane mirrors only.
[11] Carr G L, Chubar O, Dumas P, in *Spectrochemical Analysis Using Infrared Multichannel Detectors*, Ed. Bhargava R and Levin I W, Blackwell Publishing ISBN: 978-1-4051-2504-8
[12] Duncan W D, Williams G P, Appl. Optics **22**, 2914 (1983)
[13] Mao H K, Xu J, Bell P M, J. Geophys. Res. **91** 4673 (1986)
[14] Optical path for pressure determination: http://www.psi.ch/sls/ir/Station1_IntroEN/Optics_Pressure_DAC.jpg
[15] Albert S, Keppler Albert K, Lerch Ph, Quack M Faraday Discussions **150**, 1 (2011)
[16] Carroll L, Friedli P, Lerch Ph, Schneider J, Treyer D, Hunziker S, Stutz S, Sigg H Rev. Sci. Instrum. **82**, 063101 (2011)

Journal of Physics: Conference Series **359** (2012) 012004 doi:10.1088/1742-6596/359/1/012004

Status of the IR and THz beamlines at the Metrology Light Source

R Müller[1], A Hoehl[1], A Matschulat[1], A Serdyukov[1], G Ulm[1], J Feikes[2], M Ries[2] and G Wüstefeld[2]

[1] Physikalisch-Technische Bundesanstalt (PTB), Abbestraße 2-12, 10587 Berlin, Germany

[2] Helmholtz-Zentrum Berlin (HZB), Albert-Einstein-Straße 15, 12489 Berlin, Germany

E-mail: ralph.mueller@ptb.de

Abstract. The low-energy electron storage ring Metrology Light Source (MLS), a dedicated synchrotron radiation source, is in user operation since April 2008 at operating energies ranging from 105 MeV up to 630 MeV. It provides coherent THz radiation when it is operated in a low alpha mode with short electron bunches. At the MLS, two bending magnet beamlines optimized for the MIR and the THz spectral range are operational. We report the status of these two beamlines and present first results of the IR and THz measurements at the MLS.

1. Introduction

For more than a quarter of a century the Physikalisch-Technische Bundesanstalt (PTB) has been performing synchrotron radiation-based radiometry and metrology [1], using various electron storage rings. Since the start of operation of its own electron storage ring, the Metrology Light Source (MLS) [2-4], PTB has at hand an ideal source which can be flexibly operated, e.g. at any electron beam energy between 105 MeV and 630 MeV or at any electron beam current from 1 pA (single electron) up to 200 mA. Additionally, the MLS is the first electron storage ring worldwide designed and prepared for a special machine optical mode, -the low alpha operation mode-, for the production of high power coherent synchrotron radiation (CSR) in the THz spectral range [5]. The high THz power in combination with the high brilliance makes the MLS an interesting radiation source for many THz applications such as exploration of the linearity of THz detectors, microspectroscopy, near field microscopy, THz tomography, or THz electron paramagnetic resonance (THz-EPR) [6]. By tuning the low alpha optics the bunch length can be manipulated to the range of lower than 1 mm (3 ps). This time structure is used for the testing of fast THz detectors. Installed at the MLS are two IR/THz beamlines: (1) an infrared beamline optimized for the MIR, and (2) a dedicated THz beamline.

2. The IR and THz beamlines at the MLS

Typical IR beamlines at electron storage rings consist of an arrangement of mirrors which allows - in combination with a special port of the dipole chamber – the transport of the beam to the experiment (see Fig 1a). After all mirror reflections the σ-polarization of the electrical wave vector of the bending magnet radiation is oriented horizontally. The propagation of sub-terahertz electromagnetic waves from the source point to the experiment through a typical IR beamline is strongly affected by diffraction. This is why we decided to build a dedicated THz beamline with large extraction optics and a larger quartz window (Fig. 1b) [7]. Table 1 and 2 show the dimensions of the optical elements in the IR and THz beamline.

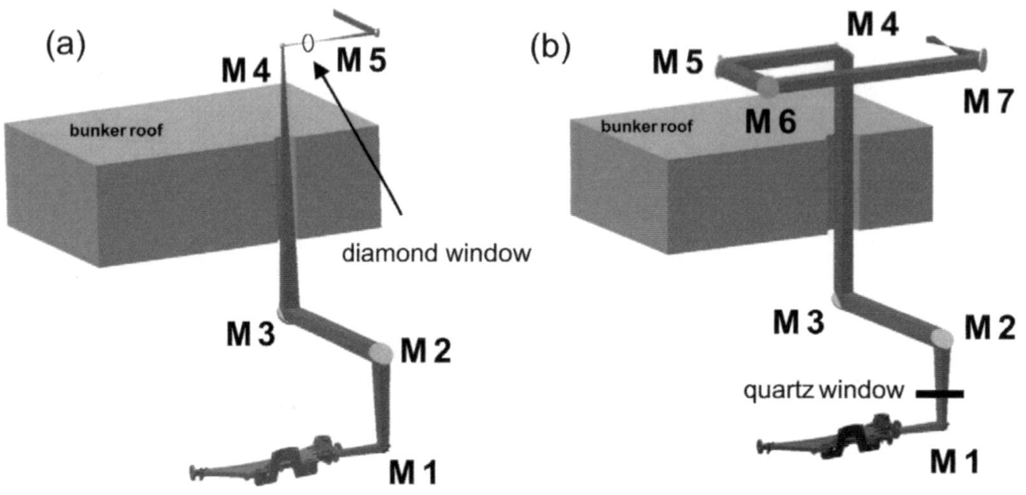

Figure 1. Layout of the IR (a) and THz beamline (b) at the MLS.

Table 1: Optical elements of the MLS IR beamline.

Opical element (mirror)	Distance to source point (mm)	Type	Dimensions (mm)
M1	1550	planar	110 × 110
M2	2450	cylindrical	130 × 240
M3	4450	cylindrical	210 × 150
M4	7050	planar	40 × 40
Window	7350	diamond	d = 30
M5	8550	planar	60 × 100

Table 2: Optical elements of the MLS THz beamline.

Opical element (mirror)	Distance to source point (mm)	Type	Dimensions d = diameter (mm)
M1	1550	cylindrical	110 × 110
Window	2072	z-cut quartz	d = 89
M2	2450	cylindrical	d = 240
M3	4450	planar	d = 240
M4	7050	planar	d = 240
M5	8450	planar	d = 240
M6	9450	panar	d = 240
M7	11850	toroidal	d = 240

2.1. IR Beamline

The experimental station at the IR beamline is equipped with an instrumentation for microspectroscopy measurements: a Fourier-transform spectrometer and an IR microscope for life and material science investigations. A state-of-the-art rapid scan FTIR spectrometer Vertex-80v (by Bruker Optics) extended for THz applications and a maximum resolution of 0.06 cm^{-1} is now operational. The infrared microscope HYPERION 3000 (Bruker Optics) completes the experimental setup.

The scheme of the optical design concept of the MLS-IR beamline is shown in Fig. 1a. The design is adapted from the concept of the IRIS beamline [8] at BESSY II, a multipurpose beamline which is equipped with two Fourier transform spectrometers, a microscope as well as an ellipsometer and provides useful IR intensities over a broad spectral range. The MLS-IR beamline is located at the bending magnet D6 at the MLS. The plane extraction mirror allows—in combination with a special port of the dipole chamber—a horizontal and vertical collecting angle of 64 mrad (h) x 43 mrad (v). At the maximum ring operating conditions of 630 MeV and 200 mA a mirror heat load of about 16 W is expected. Due to this relatively low heat load and with an additional cooling system, there is no deformation or destruction effects at the first mirror. So there is no need to split the mirror as, for example, at the IRIS beamline [8]. The first optical component M1 is placed at a distance of 1550 mm from the source, the first location possible outside the vacuum chamber of the dipole magnet. M1 deflects the photon beam upwards by 90° to a combination of mirrors which focus the beam outside the radiation shielding wall in the plane of a CVD diamond window. The second mirror M2 and the third mirror M3 focus the IR radiation vertically and horizontally, respectively. Both mirrors are cylindrical and deflect the beam by 90° towards the storage ring (M2) and upwards (M3). The beam passes the tunnel roof at a distance of 700 mm after M3. The fourth optical element M4 is a planar mirror and transports the beam to the parabolic mirror M5. M5 collimates the beam and sends it to the remaining optical system. The IR photon beam has an intermediate focus between M4 and M5 at a distance of 5800 mm from the center of M1. After all these reflections, the polarization is horizontally oriented. Ray-tracing calculations for an energy of 500 cm^{-1} indicate a focal spot size of 2.5 mm (h) · 1.0 mm (v) which should easily pass through the diamond window (30 mm opening). This focus serves as a new source point for the remaining optical system. The diamond window separates the UHV of the storage ring from the remainder of the beamline. The subsequent optical elements direct the light to the different experiments. By mounting the optics and experiments on the massive storage

Journal of Physics: Conference Series **359** (2012) 012004 doi:10.1088/1742-6596/359/1/012004

ring tunnel itself, the mechanical stability required for vibration-sensitive IR experiments can be achieved.

The IR beamline is optimized for the MIR spectral region. Figure 2 shows a comparison of the brilliance of the synchrotron light and the internal globar source of the BRUKER80v.

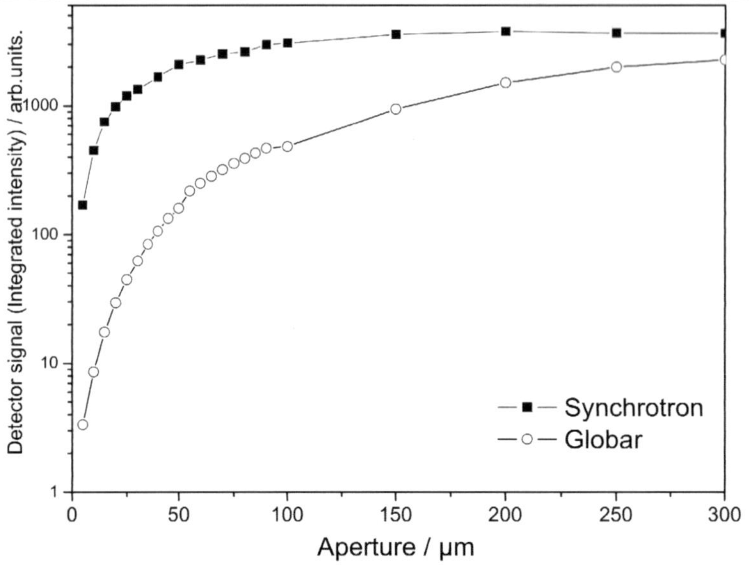

Figure 2. Comparison of the intensity of the internal globar source of the Vertex80v with the synchrotron radiation at the IR beamline after the knife-edge aperture of the IR microscope HYPERION 3000. The intensity was integrated over the wavelength range from 1.5 µm to 25 µm.

2.2 THz beamline

Figure 1b shows the design of the dedicated THz beamline. The beamline is located at the dipole D5. The acceptance of the dipole chamber and the distance between the source point and the mirror M1 are the same as for the IR beamline. The mirrors M1 and M2 collimate the THz radiation horizontally and vertically, respectively. Both mirrors are cylindrical and deflect the beam by 90° upwards (M1) and towards (M2) the storage ring. A z-cut quartz window with a diameter of 89 mm after the first mirror M1 separates the UHV of the storage ring from the rest of the beamline. The following optical elements M3 to M6 are planar mirrors and transport the beam to the toroidal mirror M7. M7 focuses the beam and directs it to the experiment. Radiation safety requires this complex optical path. The nominal diameter of the beamline tube is 250 mm throughout the THz beamline.

After mirror M7 the beam will lead to two different experimental stations: Station 1 is equipped with an FTIR spectrometer for spectroscopy and station 2 is used for the tests of THz detectors. Figure 3 shows the focus of the THz radiation (all radiation with a wavelength longer than 100 µm) in the low-alpha mode at 630 MeV at the THz beamline. Its FWHM size is approximately 1.7 mm in diameter and lies directly on the focus of the visible and near infrared light.

Journal of Physics: Conference Series **359** (2012) 012004 doi:10.1088/1742-6596/359/1/012004

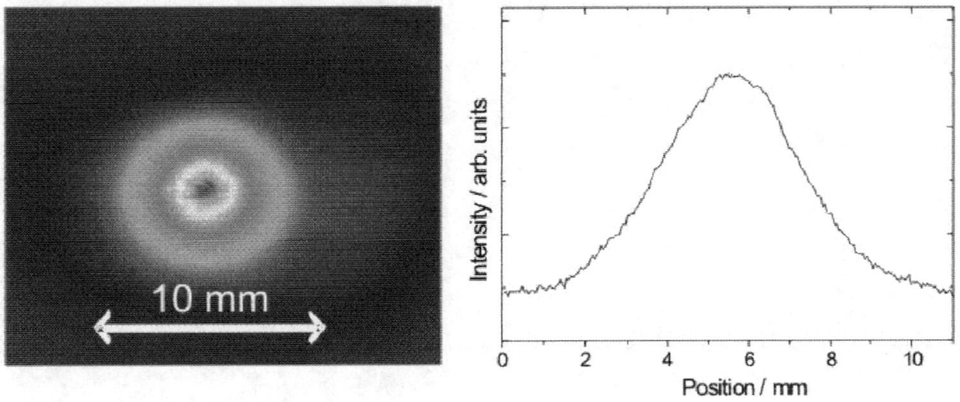

Figure 3. Left: Focus of the THz beamline measured by an infrared camera. Right: Beam profile of the focussed THz radiation.

The wavelength range at MLS, where the coherent radiation has more intensity compared to the incoherent THz spectrum, is 1.4 cm^{-1} to about 50 cm^{-1} [6]. This corresponds to a frequency range from 0.05 THz to 1.5 THz. The power gain compared to the normal operation mode is up to 4 orders of magnitude. At the IR beamline the measured power is in the range of a few hundred micro-watts. The highest measured average THz power of about 60 mW gives with the machine parameters of the MLS a peak power of about 35 W [6].

3. THz measurements

The Metrology Light Source (MLS) is the first electron storage ring worldwide designed and prepared for generation of THz radiation by applying a special machine optical mode (low-α operation mode) [5]. In this mode high power coherent synchrotron radiation (CSR) in the THz spectral range can be produced. Under normal operation of the MLS at 630 MeV, high ring currents, and bunches of about 7 mm length (1 σ bunch length) the measured far IR power is temporally smooth and varies linearly with beam current, as expected for incoherent synchrotron radiation. When the bunch length is shortened, bursts of radiation are emitted The time structure is rather complex and varies with operating conditions. At MLS the bunch length can be adjusted by varying different parameters like α value, ring current, and cavity voltage [5]. For a fixed rf-voltage the bunch length is proportional to $\alpha^{1/2}$, where α is the momentum compaction factor. By lowering α, the bunches become shorter. The MLS has a unique possibility, to control the higher orders of α and to achieve bunch length reductions by more than a factor 10 in the sub-mm range. The higher orders of α are controlled by suitably placed sextupoles and octupoles. In the following we show two examples of our current THz work: (1) A comparison of different low-alpha optical modes, and (2) first results from our water absorption measurements.

3.1 Low-alpha modes for 450 MeV and 630 MeV electron energy

At the MLS two low-α operation optic modes for different ring energies, 450 MeV and 630 MeV were prepared. We measured the absolute average CSR THz power for the low-α modes at 450 MeV and 630 MeV in the focus of the THz beamline using a Thomas Keating power meter. Figure 4 shows that the CSR THz power for both low-α modes is nearly identical.

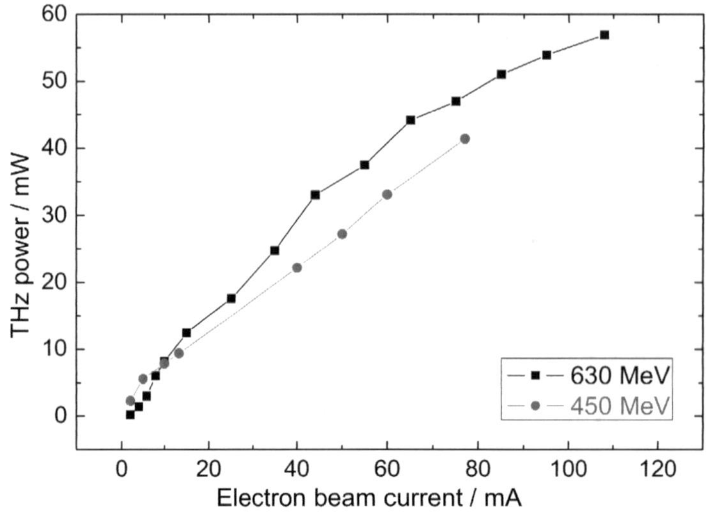

Figure 4. Comparison of the average CSR THz power measured in the focus of the THz beamline for the low-α modes at 630 MeV (squares) and 450 MeV (circles).

Figure 5 compares low-alpha spectra for 450 MeV and 630 MeV for two different electron beam currents. For both low-alpha modes the behavior is as expected: the waist and the intensity of the spectra growth with increasing electron beam current. The shape of the spectra for the different electron beam energies is comparable. From theory it is predicted that the THz radiation at the MLS is independent from the electron energy. At the MLS we find that at 105 MeV the CSR is strongly suppressed and at 250 MeV we derived a more intricate picture [9], where we saw an unexpected suppression of CSR THz signals at short bunches and high rf-voltage. So, at the MLS the low-α optics for lower electron energies is still under investigation

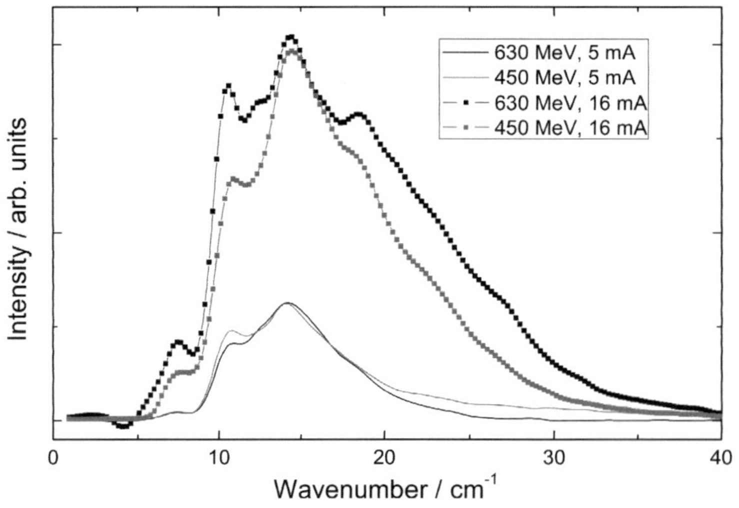

Figure 5. FTIR CSR THz spectra for the two low-α modes at 450 MeV and 630 MeV for two different electron beam currents.

3.2 THz absorption of water

The FIR or THz absorption spectrum (5 cm^{-1} to 220 cm^{-1}) of liquid water is characterized by a broad peak at 170 cm^{-1} and possibly a shoulder at 80 cm^{-1} [10, 11]. Thus it came as a surprise that a recent publication [12] of the temperature dependance of liquid water optical constants in 2009 in the temperature range of 10 °C to 50 °C (see Fig 7 of ref [12]) showed the existence of an "isosbestic point" near 12 cm^{-1} (0.36 THz) without any further discussions.. An isosbestic point refers to the frequency at which a series of absorption spectra cross i.e. frequency where the absorbance is a constant. The earlier studies reported by Afsar [10] and Hasted [11] did not indicate the presence of an isosbestic point near 12 cm^{-1} contrary to the spectra published by J.H. Son in 2009 [12]. Also the recent studies on water with the time-domain THz spectroscopy did not indicate the presence of the isosbestic point [13]. The presence of an isosbestic point in the absorption spectrum, if verified, is generally understood as the result of inter-converting distinct chemical or structural species. Alternatively, an isosbestic point in the absorption spectrum can also result from a single solute species in a fluctuating environment. In the present study, we report on absolute measurements of the liquid water transmission in the temperature range of 5 °C to 45 °C using several photon sources (coherent synchrotron radiation, incoherent synchrotron radiation, mercury-vapor lamp and globar source) to verify the existence or the non-existence of an isosbestic point near 12 cm^{-1}. We show the data "as is" without any corrections and processing. For each line we performed 15 measurements each with 128 scans and averaged on all measurements.We performed two sets of measurements: (i) with the mercury-vapor lamp in the spectral region from 5 cm^{-1} to 25 cm^{-1} and (ii) with coherent synchrotron radiation (CSR) from 3 cm^{-1} to 15 cm^{-1}. Due to the specifics of the FTIR measurements with respect to the beamsplitter transmission profile, and spectral intensity profiles of the light sources, we define the reliable spectral range for the mercury-vapor lamp from 9 cm^{-1} to 22 cm^{-1} and for CSR from 5 cm^{-1} to 13 cm^{-1}. Good agreement between the two data sets is obtained in the spectral range from 9 cm^{-1} to 13 cm^{-1} (see Fig. 6). The transmission curves for different temperatures are well separated in intensity over the whole measured spectral range, which disproves the isosbestic point existence at 12 cm^{-1}.One can see the wiggles in the spectra, which are mainly caused by the interference on the Si windows. Below circa 8 cm^{-1} the wiggles are not uniform, which lets conclude, that these structures have also another origin than the interference. Despite these disturbing effects the stability of our measurements is significantly better than ±0.3% (Fig. 6).

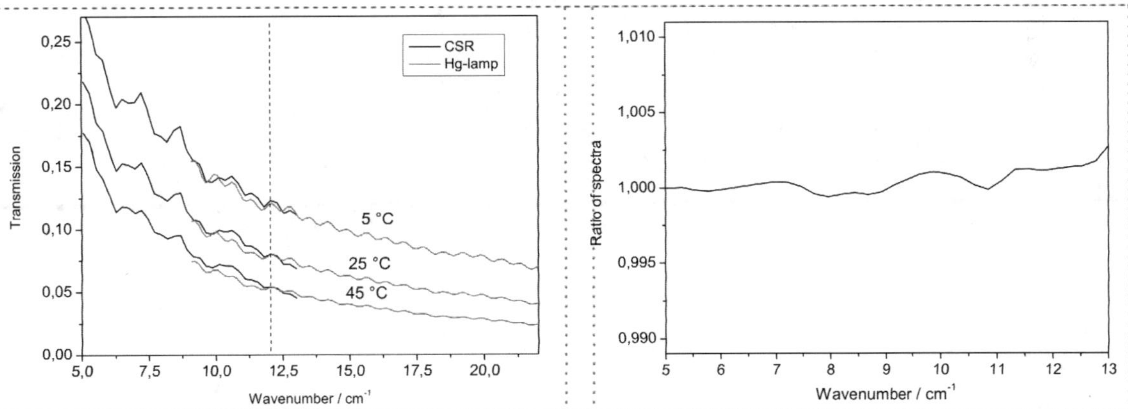

Figure 6. Left: Spectrum of the 100 µm water layer in the liquid cell (with 2 mm thick Si windows and 100 µm thick Teflon spacer), measured at different temperatures (T = 5 °C, 25 °C and 45 °C measured during warming up cycle) with the internal Hg lamp (red) and the coherent synchrotron radiation source (CSR–black). The dashed line indicates the position of the supposed isosbestic point

Journal of Physics: Conference Series **359** (2012) 012004 doi:10.1088/1742-6596/359/1/012004

at 12 cm^{-1}. The measurements were performed with the FTIR spectrometer (Bruker Optics Vertex-80v) with 0.5 cm^{-1} spectral resolution.
Right: Stability measurements of the CSR source through the fluid cell with 100 µm water sample at 15 mA ring current. To obtain this line, we took the ratio of two sequentially measured spectra (128 scans each) and calculated the average of all the measurements in the temperature series.

4. Summary and outlook

The MLS as one of the few low-energy electron storage rings worldwide is an ideal IR synchrotron radiation source. A special mode of operation allows the production of CSR and thus the production of far-IR/THz radiation with enhanced power making the MLS a promising radiation source for all types of THz applications. All beamlines are operational and equipped with a dedicated experimental set-up. The low-α mode is prepared for two different ring energies. For 450 MeV and 630 MeV THz power and spectral behavior are nearly identical and ready for user operation. Additionally, we compared the spectra of liquid water in the spectral range from 5 cm^{-1} to 22 cm^{-1} measured with two light sources – mercury-vapor lamp and CSR. Our goals for the near future are to improve the stability of the light source (CSR), and to expand the spectral region toward long wavelengths ideally down to 2 cm^{-1}.

References

[1] Beckhoff B *et al* 2009 *Phys. Status Solidi* B **246** 1415
[2] Klein R *et al* 2008 *Phys. Rev. ST Accel. Beams* **11** 110701
[3] Klein R, Brandt G, Fliegauf R, Hoehl A, Müller R, Thornagel R and Ulm G 2009 *Metrologia* **46** 266
[4] Klein R, Thornagel R, Ulm G, Feikes J, Wüstefeld G 2011 *J. Electron Spectr. Relat. Phenomena* **184** 331
[5] Feikes J, Hartrott M v, Ries M, Schmidt P, Wüstefeld G, Hoehl A, Klein R, Müller R and Ulm G 2011 *Phys. Rev. ST Accel. Beams* **14** 030705
[6] Müller R, Hoehl A, Serdyukov A, Ulm G, Feikes J, Wüstefeld G 2011 *Journal of Infrared* Millimeter and Terahertz Waves **32** 742
[7] Müller R, Hoehl A, Klein R, Schade U, Ulm G , Holldack K, Wüstefeld G 2006 *Infrared Physics & Technology* **49** 161
[8] Schade U *et al* 2001 *Rev. Sci. Instrum.* **73** 1568
[9] Wüstefeld G *et al* 2010 *Proc. of IPAC2010, Kyoto (Japan)* 2508
[10] Afsar M N, Hasted J B 1978 *Infrared Phys* **18** 835
[11] Hasted J B, Husain S K, Frescura F, Birch J R 1987 *Infrared Phys* **27** 11
[12] Son J H *J. Appl. Phys.* 2009 **105** 102033
[13] Ronne C *et al J. Chem. Phys.* 1997 **107** 5319

Polarization-modulated synchrotron infrared microspectroscopy for the study of crystalline morphology in some semicrystalline polyolefins

G Santoro[1], M Schmidt[2], U Schade[3], C Marco[1], G Ellis[1,4]
[1] Instituto de Ciencia y Tecnología de Polímeros, ICTP – CSIC, c/ Juan de la Cierva 3, 28006 Madrid, Spain
[2] Energy Biosciences Institute, University of California, Berkeley, California 94720, USA
[3] Helmholtz-Zentrum Berlin – BESSY II, Albert Einstein Str. 15, 12489 Berlin, Germany

E-mail: gary@ictp.csic.es

Abstract. The spatially resolved dichroic behaviour of a series of isothermally crystallised samples of isotactic polypropylene and polybutene-1 was investigated using polarization-modulated synchrotron infrared microspectroscopy (PM-SIRMS). The high brightness of the synchrotron source provides a good signal-to-noise performance at high spatial resolution in spite of the significant loss in infrared intensity through the photoelastic modulator. This technique gives access to information on average polymer chain orientation and crystalline morphology at the microscale, widening the possibilities for spatially resolved dichroic mapping of polymeric materials.

1. Introduction

The singular nature of the synchrotron IR radiation (coherence, high brightness, etc.) has provided important advantages over thermal sources for the study of a diverse range of problems in polymeric materials, including polymer blends and composites [1,2], multi-layered polymers [3,4], micro-structured biopolymers [5] and the crystalline morphology of semicrystalline polymers [2,6,7], amongst others. Further, the high intrinsic quasi-linear polarization bending magnet radiation can provide an additional advantage when very high spatial contrast is required in the study of anisotropic materials, allowing measurements to be made without the need for polarizing elements to be inserted in the beam path [8]. By careful alignment of a defined sample axis with the polarization vector of the synchrotron beam we have employed this approach very successfully to study the nature of polymorphic interphases in model fibre-reinforced polymer composites [6,7].

The possibility of mapping the dichroic behaviour of polymer systems is of great interest, since the degree and distribution of molecular chain orientation can provide important insight into the structure-property relations of the materials. However, one of the disadvantages for high spatial resolution SIRMS measurements is the need to carefully rotate and position the sample in order to obtain viable dichroic images. Polarization modulation provides an interesting solution since it employs a photoelastic modulator to vary the electric vector from parallel to perpendicular polarization at high

[4] To whom correspondence should be addressed.

frequency. This technique has been combined with a synchrotron source at the IRIS beamline at BESSY II Synchrotron, Berlin [9], and used effectively to measure anisotropy in inorganic crystals [10] and hydrogen bonding and orientation in wood fibres [11]. The high brightness of the synchrotron source compensates sufficiently for the loss through the additional optical elements.

In this work we describe the application of PM-SIRMS to the study of a series of carefully crystallized samples of isotactic polybutene-1 and isotactic polypropylene with different morphologies.

2. Materials and Methods

Materials: Isotactic polypropylene, iPP (Repsol-YPF, Móstoles Spain) with an isotacticity index of approximately 95% determined by NMR and a molecular weight M_v = 164,000 was employed. Film samples of around 20 μm thickness were prepared from the melt at 210 ºC in a Collin Hot Press. Subsequently, a specific thermal history was applied. The samples were heated to 210 ºC for 10 minutes in a Mettler FP-90HT hot-stage to erase the previous thermal history, and then cooled rapidly to an isothermal crystallization temperature, T_i (specified) whilst controlling the morphology using a Reichert Zetopan-Pol polarised optical microscope. Fibre-reinforced iPP model composites were also prepared by embedding carefully separated single 23 μm diameter filaments of Vectra^{TM} A950 167 Tex polyester liquid crystal polymer fibres (Ticona, Shelby, USA) between two thin-films of iPP, by sandwiching them together between glass slides above the melting temperature on a hot stage. The composites were subsequently heated to 210 ºC for 10 minutes, and cooled rapidly to an isothermal crystallisation temperature, T_i (specified in the text). At this temperature a specific mechanical history was applied by pulling the fibre using the methodology described in detail in reference [6] to generate a series of polymorphic interphase morphologies along the embedded fibre in the iPP matrix.

Isotactic polybutene-1, iPB1 (Basell, Louvain-la-Neuve, Belgium) with an isotacticity index of around 83% and a molecular weight M_n = 125,300 was employed. Thin film samples of around 20 μm were melted at 160ºC in a hot press, and then transferred to an oven held at 90 ºC for isothermal crystallization. Under these conditions relatively large spherulitic morphologies were obtained. The samples were then left to age for over 2 weeks to assure complete transformation from the tetragonal crystal modification, denominated Form II, to the hexagonal crystalline modification, Form I [12].

Methods: Synchrotron infrared microspectroscopy (SIRMS) measurements were performed at the IRIS beamline at the BESSY II storage ring of the Helmholtz Zentrum Berlin, employing a Thermo Scientific Continuμm Infrared microscope with a 32x Schwartzchild objective (NA = 0.6) and a liquid nitrogen cooled HgCdTe detector, coupled to a Nexus 870 FTIR spectrometer (Thermo Scientific).

The experimental arrangement for polarization modulation has been described in some detail previously [9] and consists of a KRS-5 holographic aluminium grid polarizer (Specac Ltd) and a PEM-90 Model II ZnSe photoelastic modulator (PEM) (Hinds Instruments) operating at 50 kHz placed between the spectrometer and the microscope, and a SSD 100 synchronous sampling demodulator (GWC Technologies) for modulation and demodulation control. When plane polarized light is incident at an angle of 45º to the optical axis of the PEM that is operated at a peak retardation of λ/2, the light beam is modulated between two perpendicular linearly polarized states at its second harmonic, 100 kHz allowing the measurement of the differential absorption between orthogonal states of linearly polarized IR light, and generates two spectra at each measurement. These correspond to the difference spectrum, $I_{diff} = I_{||} - I_{\perp}$ and the sum spectrum, $I_{sum} = I_{||} + I_{\perp}$, where $I_{||}$ and $I_{\perp}$ are the detected spectral signals for IR radiation polarized parallel and perpendicular, respectively, to the reference axis direction that corresponds with the horizontal or *x*-axis plane of the microscope stage. The ratio I_{diff} / I_{sum} yields the normalized difference spectrum, termed here as the polarization modulation dichroic difference (PMDD) spectrum. An additional advantage of the technique is that since both signals are acquired simultaneously at the same sample position, any instrumental fluctuations or atmospheric changes over time are automatically corrected for, eliminating the need for purging the sample area for example.

For both point spectra and mapping between 64 and 256 scans were recorded at each data point at a spectral resolution of 8 cm^{-1}. Various aperture sizes between 4 x 4 μm^2 and 15 x 15 μm^2 were employed for different samples. Whilst it was possible to obtain spectra at very small apertures, the resulting signal-to-noise ratios in the PMDD spectra were poor and long sampling times were required. A compromise was effectively made as a function of the spectral region used in the analysis and the overall sampling time employed, and the aperture sizes are given in the Figures. It is important to note that with a globar source PM technique is not accessible at apertures < 25 x 25 μm^2 due to the considerable energy loss through the photoelastic modulator.

3. Results and Discussion

The sensitivity of polarised infrared radiation to the orientation in polymers has for many years provided a very important tool for scientists and technologists alike, and the application of vibrational linear dichroism to study polymer chain orientation is routine, informative and highly documented. The application of linear dichroism in infrared microspectroscopy is somewhat more recent and its impact has been limited fundamentally by the lower boundary of the aperture-delimiting signal-to-noise performance that conditions the spatial contrast available. Synchrotron IR sources have been demonstrated to provide new opportunities for the study of orientation and conformational characteristics of polymeric materials using the inherently polarised source radiation, and spectra have been shown to retain orientation sensitivity even through apertures as small as 3 x 3 μm^2 [6]. However, in order to obtain dichroic information from a sample, polarised spectra both parallel and perpendicular to the experimental reference frame must be obtained at the same point on the sample. With a synchrotron source this requires that the sample be carefully rotated if the maximum polarised flux is to be retained, which inevitably leads to experimental difficulties especially when precise area or line mapping is required. The principal advantage of the polarisation modulation experiment is that it directly provides a dichroic spectrum from each data point.

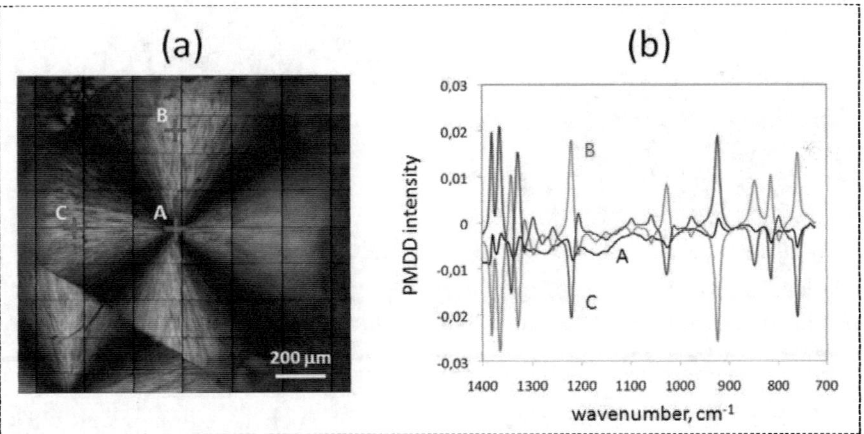

Figure 1. (a) Spherulitic morphology in iPB1 and (b) PMDD spectra recorded at the positions marked. Aperture = 10 x 10 μm^2

It is well known that polymer crystallization initiates at nucleation sites in the molten polymer matrix and if unperturbed the crystals grow in a three-dimensional manner to form semicrystalline superstructures known as spherulites [12]. Figure 1a shows an example of spherulitic crystalline morphology in a thin film sample of polybutene-1 visualised using polarised optical microscopy. The polarisation modulated dichroic difference (PMDD) spectra, recorded at 3 different points marked A, B and C on the large spherulite in the centre of the image, are given in Figure 1b.

The differences between the PMDD spectra arise from the different mean chain orientation in the region of the sample defined by the aperture. The direction of the polymer chains in a chain folded crystal structure is orthogonal to the crystal growth axis that is in the radial direction, and during crystallisation the distance from the nucleus of the growing face of the polymer crystals increases with time. Consequently, at the centre of the spherulite, position A, the lamellar crystals are arranged radially from the nucleus in all directions, almost averaging out the chain orientation in the region specified by the aperture. Whereas in the PMDD spectra at positions B and C the average molecular chain orientation is horizontal and vertical respectively, and since the distance from the spherulitic nucleus is approximately the same in both cases the spectra appear symmetrically inverted, as is expected since their angular relation in space with respect to the chain axis is also rotated 90º. This is a very clear demonstration of the high sensitivity of the technique to polymer chain orientation.

In the case of isotactic polypropylene, when it crystallises from the melt in the presence of a reinforcing fibre, the fibre surface nucleates the crystallisation, but the very high density of nucleation sites on the fibre surface restricts the freedom of growth of the growing polymer crystals such that they grow orthogonally to the fibre axis with the molecular chain axis parallel to the fibre, termed transcrystalline morphology [12]. In quiescent conditions the α-transcrystalline, α-TC form of iPP is generated, coexisting with spherulitic α-iPP that is nucleated in the matrix. However, if the matrix is subjected to shear in the presence of a fibre a more complex polymorphic morphology develops where a thin layer of α-row nuclei develop that induce a β-cylindritic morphology where the polymer chain direction is also parallel to the fibre axis. This polymorphic interphase structure has been described morphologically [13,14] and, more recently, structurally using synchrotron IR microspectroscopy and x-ray microdiffraction [6,7].

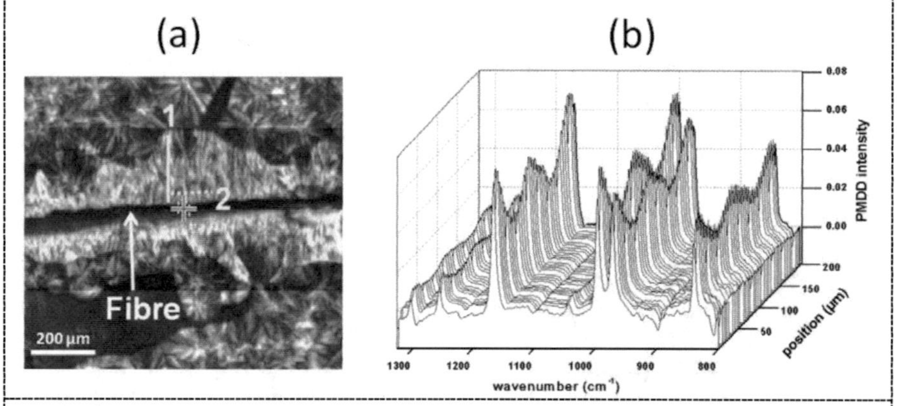

Figure 2. (a) PLM image of polymorphic region generated around a fibre in iPP and (b) PMDD spectra obtained along line 1 in Figure 2a, Aperture = 15 x 15 μm^2, step size = 5 μm.

In Figure 2a a polarised optical microphotograph of such a region in an iPP / LCP fibre-reinforced film isothermally crystallized at 130 ºC is shown, and PMDD spectra obtained in a line perpendicular to the fibre axis, line 1, using a 5 μm step size are given in Figure 2b. The periodic oscillations observed are thought to originate from the phenomenon of lamellar twisting.This is further illustrated in Figure 3 in a graph of the PMDD intensity for a mixed vibrational mode that appears at around 997 cm^{-1} *vs.* distance from the fibre. This band is highly characteristic of the crystalline phase and arises from a combination of γ_{CH3}, $\nu_{C\text{-}CH3}$, δ_{CH} and τ_{CH2} vibrations with a average dipole moment approximately parallel to the molecular chain axis,

Lamellar twisting has been observed morphologically in a number of polymer spherulites such as polyethylene [15-17], amongst others [18], and also in transcrystalline polyethylene reinforced with aramid or UHMWPE fibres [15, 19] and in iPP reinforced with Kevlar-29 fibres [20]. However, this phenomenon has not been previously reported for the β-cylindritic morphology.

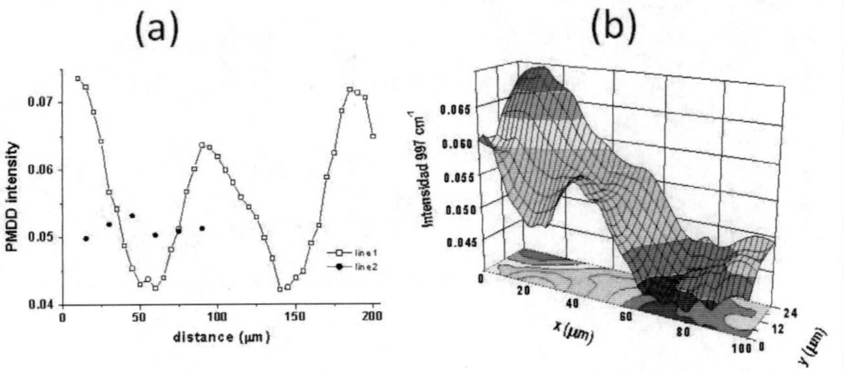

Figure 3. (a) PMDD intensity of 997 cm^{-1} band *vs* distance for the lines indicated in Fig 2a; aperture = 15 x 15 μm^2, step size = 5 µm, and (b) PMDD intensity of 997 cm^{-1} band over an area of a second sample (described in the text); aperture 12 x 12 μm^2, step size 4 µm in *x-axis* and 8 µm in *y-axis*.

Further, it is interesting to note that when spectra were obtained in a line parallel to the fibre axis, line 2 in Figure 3a, only small variations were observed in the PMDD intensity suggesting that the twisting of adjacent crystals in discrete regions along the fibre may exhibit certain cooperativity in their growth habit. This is not unreasonable if the nucleation process at adjacent sites is simultaneous. Figure 3b shows the intensity of the 997 cm^{-1} band from an area mapped using a 12 x 12 μm^2 aperture from another sample that was crystallized isothermally at 130 °C and the fibre pulled at a slightly faster rate at a temperature 2°C above T_i. Whilst evidence for the periodic oscillations observed in Figure 3a are also apparent, it can be seen that in general the twisting phenomenon diminishes with distance from the fibre, and that adjacent regions do not appear to follow the same twist periodicity. This suggests that the structure of the β-cylindritic layer is highly influenced by the shear conditions. It is clear that the control of shear along the fibre is very complex and the density of nucleation sites on the fibre surface can vary significantly.

Finally, PM-SIRMS was employed to study a polymorphic region in a sample crystallized isothermally at 142 °C for 6 hours and then cooled dynamically to room temperature at 20 °C min^{-1}. Figure 4a shows a polarized light micrograph of the sample where the region mapped through a 12 x 12 μm^2 aperture is drawn, and Figure 4b shows the spectra obtained along the dotted line shown and includes a linescan image where three separate spectral regions can be clearly observed and have been delimited with dotted lines. In Figure 4c an axiomatic projection map of the area marked by the red box in Figure 4a shows that PM-SIRMS has the capacity to clearly discriminate between these different crystalline morphologies. Figure 4d shows the PMDD spectra obtained from points I, II and III (marked in Figure 4a). Analysis of the spectral information allows us to identify the nature of the three regions: Region I corresponds to β-iPP, region II corresponds to α-iPP, and region III is most likely due to a mixed αβ-morphology favoured under these specific crystallization conditions.

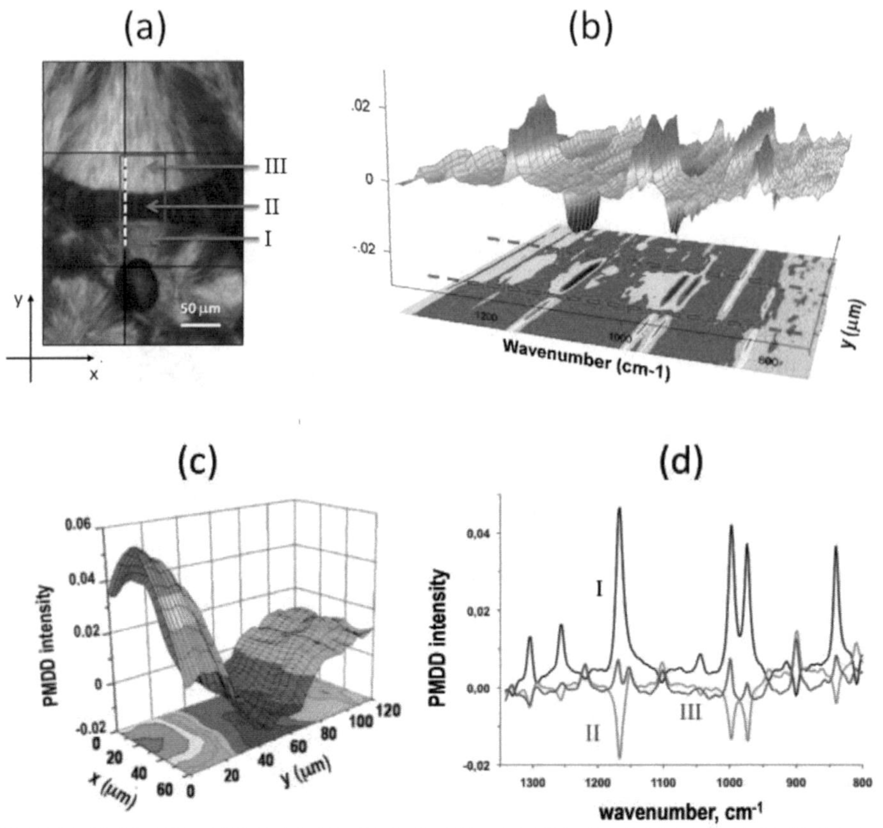

Figure 4. (a) PLM image of polymorphic region of iPP crystallized at 142°C, (b) PMDD spectra and linescan image along dotted line, (c) Mapping of PMDD intensity of 997 cm^{-1} band over the area shown by red box, and (d) spectra from the points marked. Aperture = 12 x 12 μm^2, step size = 4 μm in *x-axis* and 8 μm in *y-axis*.

4. Conclusions

Polarization modulated synchrotron infrared microspectroscopy can be a very useful tool for the study of the nature of the crystalline morphology in semicrystalline polymers and their composites. We have shown that the PMDD spectra are highly sensitive to polymer chain orientation, and the high spatial resolution allows us to obtain good quality spectra from within polymer spherulites, in the case of both isotactic polypropylene and isotactic polybutene-1, and in transcrystalline morphologies and polymorphic interphases in isotactic polypropylene. In the case of iPP, periodic variations in band intensities in the PMDD spectra were presented as evidence for lamellar twisting, which depends on the crystallization conditions of the polymer matrix, observed for the first time in the β-cylindritic morphology of iPP. Future work is underway to contrast this data with other techniques such as electron microscopy and, in particular, x-ray microdiffraction, and to deepen the knowledge that infrared microspectroscopy can provide on the variables that control the nature of both spherulitic and transcrystalline structures. Finally, the brightness of the synchrotron source is fundamental since such high spatial resolution polarization modulation measurements are not accessible with a *globar* source. The PM-SIRMS technique thus becomes a powerful addition to the spectroscopists' toolkit for the characterisation of anisotropic order at the microscale.

5. Acknowledgements
The authors wish to acknowledge the financial support from the Spanish Ministry of Science and Innovation (MICINN) Project MAT-2010-21070-C02-01, and from the European Community's Seventh Framework Programme (FP7/2007-2013) under grant agreement n.°226716. GS wishes to thank the MICINN for a FPI Research Contract.

6. References
[1] Ellis G, Marco C, Gómez M A, Collar E P and García-Martínez J M 2004 *J. Macromol. Sci. Phys.* **43B** 253
[2] Ellis G, Marco C and Gómez M 2004 *Infrared Phys. Technol.* **45** 349
[3] Sankhe SY and Hirt DE 2002 *Appl. Spectrosc.* **56** 205
[4] Vernoud L, Betchel H A, Martin M C, Reffner J A and Blackledge R D 2011 *Forensic Sci. Int.* **210** 47
[5] Ellis G, Cano P, Jadraque M, Martin M, López L Nuñez T, De La Peña E, Marco C and Garrido L 2011 *Anal. Bioanal. Chem.* **399** 2379
[6] Ellis G, Marco C and Gómez M A 2004 *J. Macromol. Sci. Phys.* **43B** 191
[7] Torre J, Cortazar M, Gómez M A, Marco C, Ellis G, Riekel C and Dumas P 2006 *Macromolecules* **39** 5564
[8] Santoro G, Yousef I, Jamme F, Dumas P, Ellis G 2011 *Rev. Sci. Instrum.* **82** 033710
[9] Schmidt M, Schade U and Grunze M 2006 *Infrared Phys. Technol.* **49** 69
[10] Schmidt M, Lee JS, Grunze M, Kim KH and Schade U 2008 *Appl. Spectrosc.* **62** 171
[11] Schmidt M, Gierlinger N, Schade U, Rogge T and Grunze M 2006 *Biopolymers* **83** 54
[12] Basset DC "Principles of Polymer Morphology" 1981 Cambridge Univ. Press, Cambridge, UK
[13] Varga J, Karger-Kocsis J 1996 *J. Polym. Sci. Part B – Polym. Phys.* **34** 657
[14] Wu CM, Chen M, Karger-Kocsis J 1998 *Polym. Bull.* **41** 493
[15] Keith HD and Padden FJ 1984 *Polymer* **25**, 28
[16] Kunze M, Dreschler M and Möller S 1995 *Polymer* **36** 1331
[17] Assouline E, Watchel E, Grigull S, Lustiger A, Wagner HD and Marom G 2001 *Polymer* **42** 6231
[18] Lotz B and Cheng SZD 2005 *Polymer* **46** 577
[19] Ratner S, Mona Moret P, Watchel E and Marom G *Macromol Chem Phys* **206** 1183
[20] Dean DM, Rebenfeld L, Register A and Hsiao B 1998 *J Mat Sci* **33** 4797

Journal of Physics: Conference Series **359** (2012) 012006 doi:10.1088/1742-6596/359/1/012006

Optical standing-wave artifacts in reflection-absorption FTIR microspectroscopy of biological materials

J Filik, M D Frogley[1], J Pijanka, K Wehbe and G Cinque
Diamond Light Source Limited, Didcot, Oxfordshire, OX11 0DE, UK.

E-mail: mark.frogley@diamond.ac.uk

Abstract. Reflection-absorption spectra obtained with an infrared microscope should yield the same absorption coefficients as direct micro-transmission measurements as long as the correct effective sample thickness is used, but in practice, severe optical artifacts can complicate the spectra. Using deposited protein gel films as a homogenous model for biological cell-like samples, we demonstrate the effect of standing-wave interference of the IR beam at the reflective substrate surface which dramatically and systematically alters the absorbance intensity across the spectrum as a function of sample thickness. To explain the observed spectral artifacts, we simulate the optical standing-wave for the focussed IR beam, and insert the parameters into an existing standing-wave absorption theory. By introducing an additional term to the theory representing a component of the standing-wave resonant with the film thickness, the data are accurately reproduced, and the relative band intensities can be corrected to the direct transmission values. This approach may be generally applicable in reflection-absorption experiments to obtain reliable absorbance spectra of homogenous samples even when the sample thickness is larger than the IR wavelength.

1. Introduction

Reflection-absorption infrared spectroscopy (RAIRS) is an essential technique to study thin films or thin particles on a reflective substrate [1,2]. Traditionally, the infrared beam is used in grazing incidence so that the absorbance signal is greatly increased by the double pass through the sample, the increased path length and the enhanced electric field near the reflective substrate [3]. More recently, the reflection-absorption geometry has found broader application in microspectroscopy, for near-normal incidence 'transflectance' measurements on thicker samples due to the very broad spectral range, zero dispersion and low cost of the substrates compared with IR transmission windows.

The shape of reflection-absorbance spectra should be directly comparable with the absorbance calculated from transmission spectra, but several optical artifacts must be taken into account. Mie [4] and resonant-Mie [5] scattering from the sample shape and refractive index contrast introduce complex structure in the spectra including broad oscillations and sharp dispersive features which distort the baseline and absorption peaks. These features are generally more significant in the reflection geometry, and algorithms have been developed to fit and subtract the components with some success [6-8]. Reflected light from the top of the sample can interfere with that reflected from the

[1] To whom any correspondence should be addressed.

substrate causing Fabry-Perot fringes across the spectrum, and also add a dispersive artifact due to the specular reflectivity spectrum of the sample [9].

In our recent microspectroscopic measurements [10] of isolated human cells on MirrIR reflective substrates, Mie and resonant-Mie scattering was observed, but the most striking artifact was the systematic variation in band intensities across the spectrum as a function of position on the cells which was not observed in transmission measurements. This effect has been observed in *Bacillus subtilis* endospores [11] and recently in cellulose films [12] and has been largely explained [13] in a model system of a polymer-bead film by considering the optical standing-wave above the mirror surface caused by the 180° phase shift of the s-component of the incident electric field at the surface [14]. The absorbance at each wavelength is enhanced at the antinodes of the standing-wave for that wavelength and suppressed at the nodes, dropping close to zero for all wavelengths at the mirror surface.

In this work we have investigated the standing-wave effect in protein films as a model system for biological cells - having a similar thickness variation, surface shape and refractive index, but with homogenous composition. We compare these results with the previous analysis of reference 13 and show the data can only be accurately described if resonance effects are included when the sample top surface approaches an antinode of the standing-wave.

2. Experimental

2.1. Sample preparation

The protein gels were produced on IR substrates using drop-coating deposition (DCD), a common sample preconcentration method used in Raman spectroscopy. Convection currents during the evaporation of the droplet result in the formation of annular ring films with varying thickness. The final thickness and curvature of the deposited gel can be tailored by adjusting the concentration and volume of the drop. Protein samples were prepared by dissolving Bovine Serum Albumin (BSA, Sigma Aldrich) in de-ionised water (18.2 MOhm/cm) to produce a 1 mg/ml solution. Aliquots of this solution were further diluted to produce samples of concentrations between 0.5 and 0.0625 mg/ml. Rings with edge profiles of a similar size to a cell (1 μm height, 60 μm diameter) were produced using drops of 0.75 μl with a concentration of 0.5 mg/ml, (figure 1a). Samples were deposited on MirrIR (Kevley Tech., USA), and Al coated glass slides for reflectance measurements and 2 mm thick ZnS windows for transmission.

2.2. Film thickness measurement

The film thickness for each IR measurement point, typically between 0.2 and 1 μm, was accurately determined through direct measurement with an atomic force microscope (AFM, Nanonics MV1000) with a glass fibre cantilever tip in intermittent contact mode. Height profiles across BSA rings on MirrIR (figure 1b) and ZnS (figure 1c) are shown in figure 3.

2.3. Infrared absorption measurements

Reflection-absorption and transmission spectra were obtained with the synchrotron IR source (MIRIAM) at Diamond Light Source, UK, coupled to a Bruker Vertex 80V FTIR and Hyperion 3000 infrared microscope. The sampling area of 10x10 μm^2 was defined by slits in the intermediate image plane between the sample and the 50x50 μm mid-band MCT detector. 128 scans were averaged per spectrum at resolution $4 cm^{-1}$. Measurements were also taken with the conventional globar source and 20x20 μm^2 slits, where the larger slit size, 256 scans and $8 cm^{-1}$ resolution were needed to obtain good quality spectra from the thin films. For reflection-absorption measurements, the incident light filled one half of the 36x (NA = 0.5) cassegrain objective and the reflected light was collected by the other half. For direct transmission measurements a filled 36x condensor and objective matched pair was used. All other measurement parameters were kept constant.

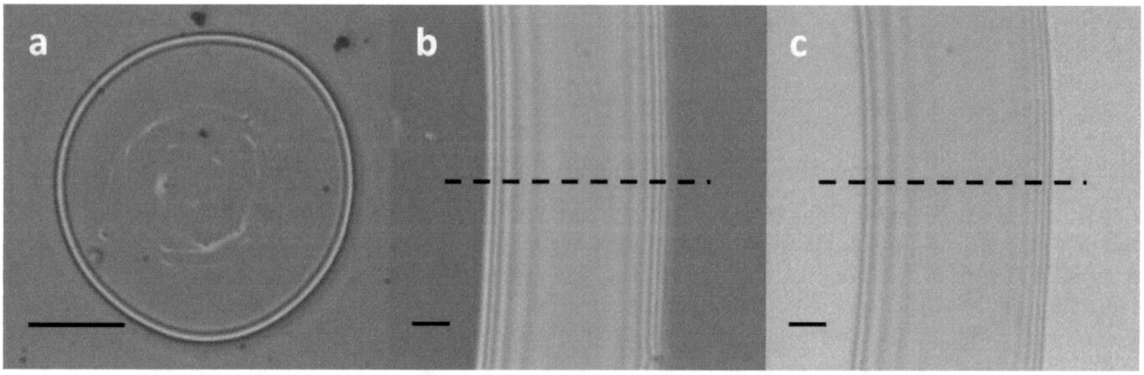

Figure 1. Visible light images of ring films of Bovine Serum Albumin. a) Ring deposited on MirrIR, scale bar 500 μm (brightfield, reflection). b) Part of the ring on MirrIR - IR measurements and height profiles were obtained across the dashed line, scale bar 10 μm. c) Ring deposited on ZnS (reflection, brightfield), scale bar 10 μm.

3. Reflection-absorption spectra of BSA films

Figure 2a) shows the reflection-absorption spectrum at different thicknesses across a BSA film. At first the high-wavenumber C-H stretch (2800-3000 cm^{-1}) and O-H/N-H stretch (3000-3700 cm^{-1}) bands increase in absorbance with thickness as expected, but for thicker regions the absorbance remains approximately constant. In contrast, the Amide I and II bands (1500-1700 cm^{-1}), which are comparable in intensity to the O-H/N-H band for thin areas, strengthen rapidly up to the thickest region of the film. The integrated intensity ratio of the C-H stretch band to the Amide (I+II) is shown in figure 4 (broken curve), and the natural interpretation of these results would be that the composition of the film varies dramatically as a function of thickness. However, when direct transmission spectra were obtained for the similar film on a ZnS window, the spectral shape (figure 2b) is essentially independent of thickness, that is, all bands increase with thickness at about the same rate and the C-H/Amide(I+II) ratio is roughly constant as expected for a film of uniform composition as shown in figure 4, solid curve.

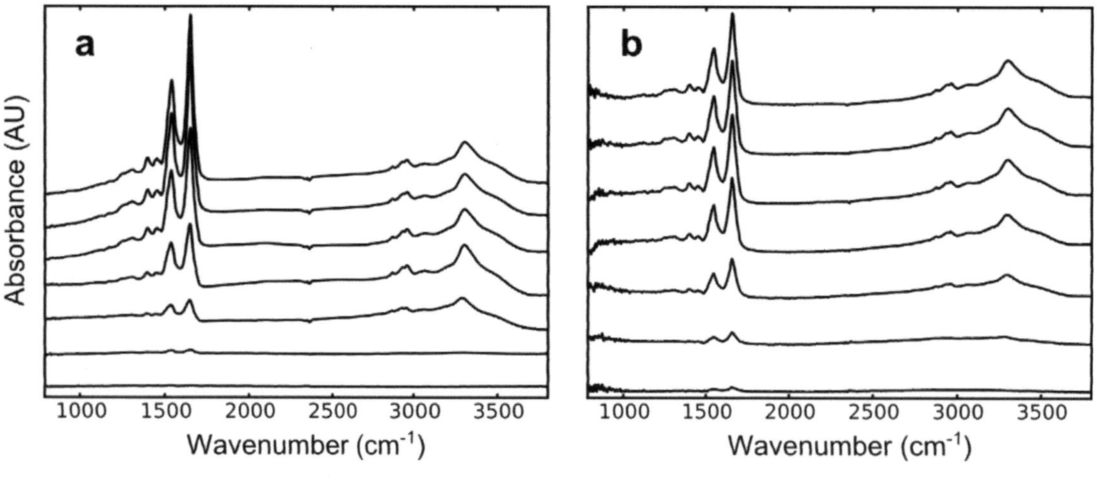

Figure 2. IR absorbance spectra for regions of different thickness in a BSA film obtained by (a) Reflection-absorption on a MirrIR substrate, film thickness from 0.26 to 0.84 μm and (b) Transmission through a ZnS substrate, thickness from 0.25 to 0.87 μm. Thickness increases for the upper curves, and each is offset for clarity.

Journal of Physics: Conference Series **359** (2012) 012006 doi:10.1088/1742-6596/359/1/012006

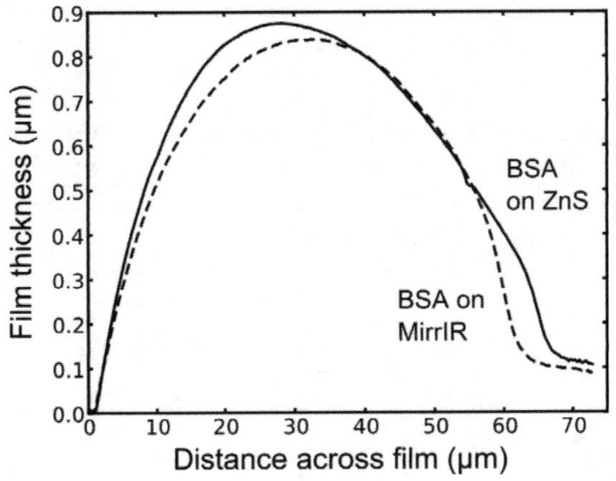

Figure 3. Thickness profiles of the BSA films measured with an atomic force microscope.

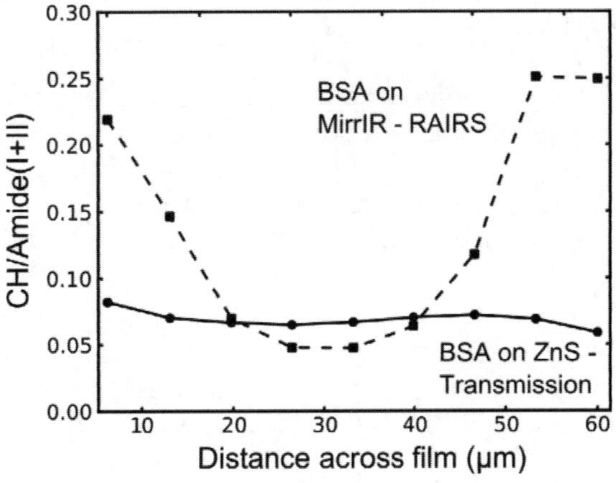

Figure 4. Ratio of the integrated CH absorbance (2800-3000 cm^{-1}) to Amide(I+II) bands (1500-1700 cm^{-1}) across the film profiles in figure 3.

4. Theory and analysis

4.1. Simulation of the standing wave

To model the standing-wave effect for unpolarised light, we performed simulations in ZEMAX optical design software, including real design parameters for the half-filled 36x cassegrain objective. Unlike previous works, this includes the effects of off-normal and highly non-collimated rays, and any optical aberrations due to the objective design and the presence of the film. For perfectly collimated incident light, the standing-wave period is $\lambda/2n$ (at normal incidence) where n is the real refractive index of the film, and *increases* with incidence angle whereas for cassegrain objectives we found the period is always less than $\lambda/2n$ and *decreases* with the average incidence angle. For refractive index values n = 1.3 to 1.8 we found a period of $0.9\lambda/2n$ with the 36x (NA = 0.5). If larger NA objectives or lower refractive index samples are used, the deviation from the collimated beam approximation is predicted to be even more significant, and the details of these simulations will be reported elsewhere. Assuming the standing wave has constant amplitude over the film thickness, the time-averaged components of the electric field as a function of height, z, above the mirror are:

$$\bar{E}_x^{\,2}(z) = A\sin^2\frac{2\pi nz}{0.9\lambda}, \qquad \bar{E}_y^{\,2}(z) = B\sin^2\frac{2\pi nz}{0.9\lambda}, \qquad \bar{E}_z^{\,2}(z) = C\cos^2\frac{2\pi nz}{0.9\lambda} \tag{1}$$

where A, B and C are constants and the dimensionless normalised intensity is:

$$I(z) = \frac{1}{(A^2+B^2)} \times \left[\bar{E}_x^{\,2}(z) + \bar{E}_y^{\,2}(z) + \bar{E}_z^{\,2}(z)\right] \tag{2}$$

4.2. Calculation of the absorbance

Following the method of reference 13, the total absorbance $a(\lambda)$ of a film of thickness l is proportional to the integral of the intensity over the standing wave:

$$a(\lambda) = a_0(\lambda)\int_0^l I(z)dz \tag{3}$$

where $a_0(\lambda)$ is a spectrum of scaling factors proportional to the spectrum of true absorption coefficients (absorbance per unit thickness).

4.3. Determination of film refractive index

In the reflection-absorption data, slow baseline oscillations are observed in all the spectra, and these may either be a result of losses due to internal reflection at the film surface or due to surface scattering as suggested in reference 13. In either case the losses depend on the intensity of the standing wave at the film surface and the expected sinusoidal contribution to the absorbance is the same:

$$F(\lambda) = D\sin^2\frac{2\pi nl}{0.9\lambda} \tag{4}$$

where again the factor 0.9 is appropriate for a 36x objective and D is a constant. Fitting equation 4 to all the spectra (e.g. figure 5) gives $n_{rms} = 1.60$ with a standard deviation $\sigma_n = 0.14$, which is consistent with other measurements on BSA films and other biological media, and was used in all the analysis.

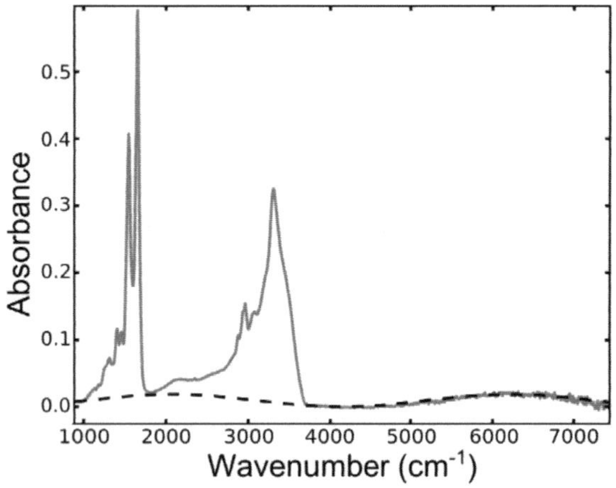

Figure 5. Oscillation in the baseline of the raw absorbance data (grey curve) is fitted with equation 4 (dashed curve) to determine the refractive index of the film.

4.4. Comparison with experiment

Figure 6 shows the measured integrated absorbance of the C-H and Amide(I+II) bands from several reflection-absorption experiments on BSA films (filled symbols) along with the results of equation 3 (broken curves). In these experiments, measurements were taken at the local highest point of the protein ring to minimise thickness variation over the sampling area, and the different heights are at different points around the ring (figure 1a). The only free parameters in the theory are the $a_0(\lambda)$ at the central wavelength of each band. For the dimensionless intensity of equation 2, only the relative values of A, B and C are important, and these were determined from the ZEMAX simulations, giving $C^2 \sim (A^2 + B^2)/40$. In practice such a small C has little effect on the results, so we set $C = 0$ and $A^2 + B^2 = 1$. For the Amide region (figure 6b), the theory (dashed curve) gives a reasonable description of the data, whereas for the C-H bands (6a), it predicts the plateau at about the right film thickness, but equation 3 is incapable of producing a shrinking absorbance with thickness. Exactly the same phenomenon was observed in the polymer bead experiments [13]. The peak and subsequent minimum in absorbance occur at film thickness of approximately $\lambda/4n$ (the first antinode of the standing-wave) and $\lambda/2n$ (the first node) respectively, and this strongly suggests a resonant contribution to the absorbance. This is reasonable since the internal reflection from the top of the film will be maximised when the antinode of the standing wave is at the surface, and the downward reflected electric field is in phase with that from the reflective substrate - there is no phase change on internal reflection. When the node is at the surface, the internal reflection is minimised and the absorbance drops. Another possible explanation is enhanced surface scattering as suggested in [13], but this could only contribute as a slowly oscillating baseline offset in the raw data (figure 5), which is effectively removed by the local linear baseline subtraction used to obtain integrated absorbance. Assuming the resonant contribution depends on the standing-wave intensity at the film surface, the enhancement would be multiplicative, and vary as $\sin^2(2\pi nl/0.9\lambda)$ and so equation 3 is modified:

$$a(\lambda) = a_0(\lambda)\int_0^l E^2(z)dz \times \left[1 + R(\lambda)\sin^2\frac{2\pi nl}{0.9\lambda}\right] \quad (5)$$

where $R(\lambda)$ is the resonance coefficient at each wavelength. For $a_0 = 1.62 \times 10^{-3}$ and $R = 1.13$, an excellent fit to the data is achieved (solid curve in Figure 6a). For the Amide bands, the same value of R with $a_0 = 0.204$ gives a notable improvement in agreement between theory and data (figure 6b). Importantly, after independently fitting the data from the two bands for these parameters, the ratio of absorbances, $a_0(\lambda_{\mathrm{CH}})/a_0(\lambda_{\mathrm{Amide}}) = 0.08$ which is in excellent agreement with the ratios from the direct transmission measurements (solid circles, figure 4). This strongly suggests that the $a_0(\lambda)$ of equation 5 are indeed proportional to the true absorption coefficients of the film. Equation 5 may therefore represent a correction function to obtain realistic absorption spectra from reflection-absorption measurements of a homogenous material where the optical thickness and $R(\lambda)$ are known, and further measurements are in progress to verify this.

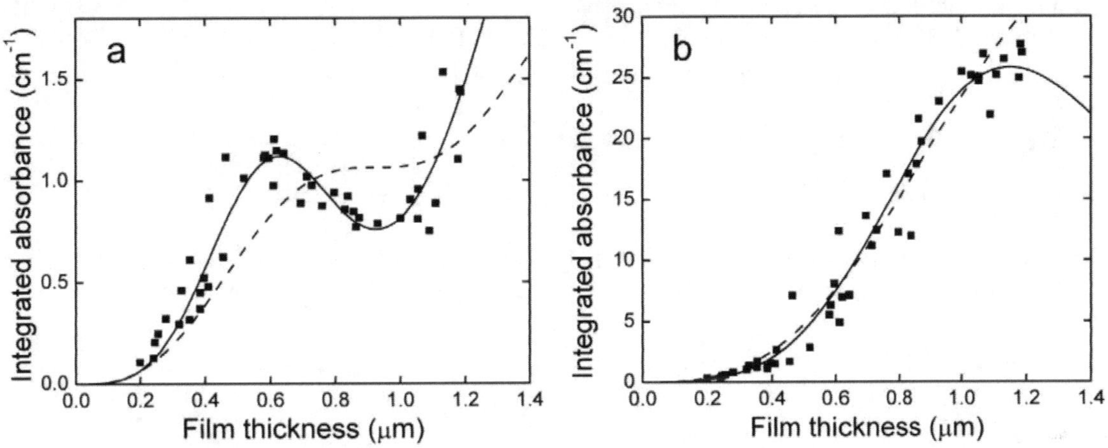

Figure 6. Integrated reflection-absorbance of (a) CH and (b) Amide(I+II) bands for several BSA films (solid squares). The broken curves are the simple standing wave theory of equation 3 and the solid curves include the resonance term, equation 5.

5. Application to inhomogenous samples

When the sample composition is not uniform, standing-wave effects are much more complex than for homogenous materials. In equation 5, $a_0(\lambda)$ and n become unknown functions of 3D position and $R(\lambda)$ cannot be directly determined. To correct the absorbance values it is then necessary to fit for the unknown parameters, for example using an algorithm similar to those developed [6-8] for scatter correction. For single cells on reflective substrates, both scattering and standing-wave artifacts are generally present at the same time and to deconvolve them, although possible in principle, presents a significant challenge.

6. Conclusion

In reflection-absorption spectra of homogenous biological films, large systematic variance of the absorption band intensities are present as a function of film thickness, which can be attributed to the optical standing-wave pattern induced by the reflective substrate. Considering only the integrated absorbance over a simple sinusoidal standing-wave is not sufficient to explain the experimental data, but by including a resonant term in the theoretical model, a good fit to the data can be achieved. The only free parameters are the resonance coefficient, which is constant for the absorption bands considered here, and a global scaling factor at each wavelength which is proportional to the sample absorbance measured in the absence of the standing-wave effect (transmission geometry). This model can assist the interpretation of spectra from inhomogenous biological samples such as single cells, but accurate correction of the artifacts will then be difficult due to the convolution of scattering and standing wave effects.

6. References

[1] Francis S A and Ellison A H 1959 J. Opt. Soc. Am. **49** 131
[2] Umemura J 2002 *Handbook of Vibrational Spectroscopy* vol 2 ed J M Chalmers and P R Griffiths (Chichester: Wiley) p982 and references therein.
[3] Greenler R G 1966 *J. Chem. Phys.* **44** 310
[4] Mie G *Ann. Phys.-Berlin* 1908 **25** 377
[5] Bassan P, Byrne H J, Bonnier F, Lee J, Dumas P and Gardner P 2009 *Analyst* **134** 1586
[6] Romeo M and Diem M 2005 Vib. Spectroscopy **38** 129
[7] Bassan P, Kohler A, Martens H, Lee J, Byrne H J, Dumas P, Gazi E, Brown M D, Clarke N W and Gardner P 2010 *Analyst* **135** 268
[8] Kohler A, Sule-Suso J, Sockalingum G D, Tobin M, Bahrami F, Yang Y, Pijanka J, Dumas P, Cotte M, van Pettius D G, Parkes G and Martens H 2008 *Appl. Spectrosc.* **62** 259
[9] Bassan P, Byrne H J, Lee J, Bonnier F, Clarke C, Dumas P, Gazi E, Brown M D, Clarke N W and Gardner P 2009 *Analyst* **134** 1171
[10] Filik J, Frogley M D, Pijanka J, Wehbe K and Cinque G 2011 Paper in preparation
[11] Brooke H, Perkins D L, Setlow B, Setlow P, Bronk B V, and Myrick M L 2008 *Appl. Spectrosc.* **62** 881
[12] Djak M, Gilli E, Kontturi E and Schennach R 2011 *Macromolecules* **44** 1775
[13] Brooke H, Bronk B V, McCutcheon J N, Morgan S L and Myrick M L 2009 *Appl. Spectrosc.* **63** 1293
[14] Wiener A Wied. Ann. 1890 **40** 203

Journal of Physics: Conference Series **359** (2012) 012007 doi:10.1088/1742-6596/359/1/012007

Synchrotron radiation infrared microspectroscopy of single living cells in microfluidic devices: advantages, disadvantages and future perspectives

L Vaccari[1], G Birarda[1,2], G Grenci[2], S Pacor[3], L Businaro[4]

[1]Sincrotrone Trieste, S. S. 14 - km 163,5, 34149 Basovizza, Trieste, Italy

[2] CNR-IOM, Laboratorio TASC- LILIT beamline, S. S. 14 - km 163,5, 34149 Basovizza, Trieste, Italy

[3] Department of Life Science, via L. Giorgieri, 34127, University of Trieste, Trieste, Italy ly

[4] CNR - Istituto di Fotonica e Nanotecnologie, Via Cineto Romano, 42, 00156 Roma, Ita

E-mail: lisa.vaccari@elettra.trieste.it

Abstract. The possibility to fully exploit the diagnostic capabilities of SR-IRMS for studying single living cells under physiological conditions is limited by several constrains. First of all, the technology for manufacturing materials transparent to both IR and visible light is quite immature, limiting the design of fluidic devices to simple demountable liquid cells. In addition, the water spectral features become prominent in the Mid IR, hiding several cellular bands and therefore limiting the diagnostic capabilities of SR-IRMS. The overcoming of the so called "water absorption barrier" requires the improvement of the protocols for the compensation of buffer spectral contributions, a goal that can be achieved also advancing the quality of IR-suitable fluidic devices. In this paper, the technical solutions employed for microfabricating completely sealed IR-visible transparent fluidic devices for living cell analysis will be presented. Several examples of the results obtained in the study of living U937 monocytes subjected to different stimuli will be selected for highlighting both the advantages and the disadvantages offered by our approach for cellular biology.

1. Introduction

In the post-genomic era many attentions have been devoted to the understanding of structure and dynamics of macromolecules, considering their multiple complex interactions as a part of a biological system, following a multidisciplinary approach known as system biology. It is therefore not surprising that in recent years several Nobel Prizes have been awarded to scientists who have further developed well established analytical techniques for the study of biological systems including mass spectrometry

Journal of Physics: Conference Series **359** (2012) 012007 doi:10.1088/1742-6596/359/1/012007

(Chemistry 2002 John B. Fenn and Koichi Tanaka), NMR spectroscopy (Chemistry 2002, Kurt Wüthrich), magnetic resonance imaging (Physiology or Medicine 2003, Paul C. Lauterbur and Sir Peter Mansfield) and fluorescence microscopy (Chemistry 2008, Osamu Shimomura, Martin Chalfie and Roger Y. Tsien). This testifies that the development of new analytical techniques as well as the more effective exploitation of existing methods is crucial for obtaining a better picture of the details of bio-molecules and/or bio-processes, part of the entangled network of the biological organization. However, the complexity of the biological systems is such that no single experimental method can provide information on all aspects of molecular structure and function, first of all because no one satisfies all the bioscience requirements, in short a label-free single-cell based assay sensitive, selective, working under physiological conditions (or better *in vivo*), no damaging and having nano-scale spatial and temporal resolution, or even better. InfraRed MicroSpectroscopy (IRMS) of cellular sample is not a single-molecule detection technique but it is a label-free method, very sensitive and not damaging, also if operated with Synchrotron Radiation (SR) [1], whose brilliance allows achieving diffraction-limited spatial resolution, which is of few microns in the MIR [2]. SR-IRMS has been largely employed for the characterization of individual single cells [3], also with sub-cellular spatial resolution, but it has been almost exclusively limited to fixed or dried cells, due to several limitations imposed by aqueous media in both sampling techniques and data analysis. These limitations can be roughly classified into technical and spectroscopic.

From a technical point of view, IR suitable materials impose many constrains on the fabrication of proper fluidic devices for living cell analysis. Recently published works on the study of living species under physiological conditions have been carried out by using commercial or custom made demountable liquid cells, produced spacing apart two optical windows by using a plastic spacer [4-6]. This approach does not ensure any design flexibility, limiting the fluidic design to a "pond" for cells, preventing any possibility to both manufacture fluidic elements, such as microchannels or microwells, and integrate other device functionalities, in a way similar to more complex Lab-On-a-Chips (LOCs). Moreover, the optical path in demountable cells cannot be carefully controlled: it is not perfectly reproducible from an experiment to another, and possibly differs in distinct device regions. All these aspects make even more difficult the overcoming of the so called "water absorption barrier".

When collecting the microspectrum of a living cell, assumed of spherical shape, setting the knife-edge apertures of Vis-IR microscope in order to fit the cell diameter, more than 80% of the sampled volume is constituted by water, ~48% of extra-cellular and ~36% of intracellular water, considering the cellular density close to 1 Kg/m^3 . It is therefore not surprising that the IR microspectrum of a living cell is dominated by water features (see Figure 1a), looking quite unusual if compared to standard fixed cell spectrum (see Figure 1b). The latter reflects the composition of a dried cell, made of ~ 50% of protein, ~ 15% of carbohydrates, ~ 15% of nucleic acids, ~ 10% of lipids and ~ 15% other molecules for an eukaryotic animal cell [7]. When the same cell is measured under physiological conditions, stretching bands of amine and hydroxyl moieties, mainly of proteins and carbohydrates, are completely hidden by the symmetric (ν_1) and asymmetric (ν_3) stretching bands of water O-H while bands characteristic of nucleic acids and carbohydrates falling below 1400 cm^{-1} as well as symmetric and asymmetric methyl and methylene stretching bands between 3000 and 2800 cm^{-1} are still discernable. For what concerns proteins, the water bending band centred at ~1643 cm^{-1} (ν_3) completely hides Amide I band and partially the Amide II, limiting the diagnostic capabilities of IRMS, especially toward cellular proteins, unless the buffer contribution to IR single-cell microspectra is compensated.

In this paper, both technical and spectroscopic aspects of single cell measurements under physiological conditions with SR-IRMS will be addressed. Selected experiments done over the past years at SISSI beamline (Synchrotron Infrared Source for Spectroscopy and Imaging) at Elettra [8] will be explained for highlighting advantages, disadvantages and future perspectives toward the upgrade of SR-IRMS to a mature label-free single cell-based assay.

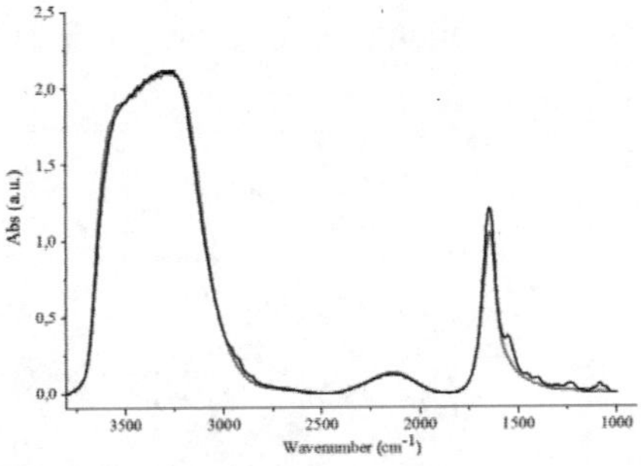

Figure 1a. SR-IR microspectra of an individual living U937 monocyte and buffer, black and red lines respectively.

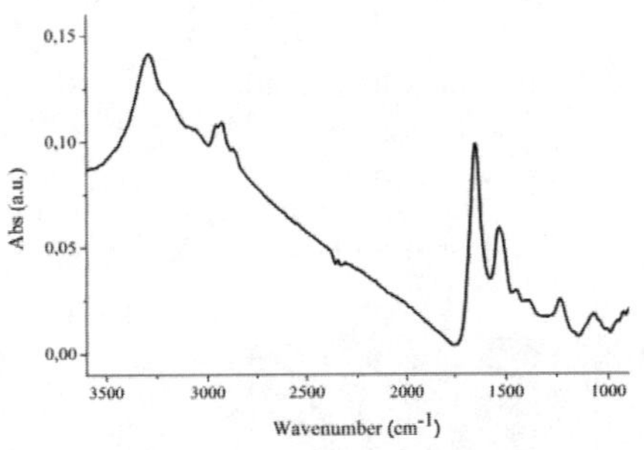

Figure 1b. SR-IR microspectrum of an individual U937 monocyte formalin-fixed.

2. Experimental session

2.1. Fluidic devices fabrication

In principle any design can be obtained by lithography, with spatial resolution of the details and fabrication time imposed by the type of lithographic approach chosen. Photolithography (also known as UV-lithography) allows to easily achieve micrometric resolution, perfectly matching the lateral resolution of SR-IRMS; it is fast and cheap, and therefore it was our first choice. As an example of device fabrication, we will detail the protocol used for manufacturing the device shown in Figure 2a. XARP 3100/10 (All Resist GmbH, DE) photoresist was spun onto a 2 mm thick CaF_2 optical window 3 cm in diameter. The achieved resist thickness was 7,5 microns. The resist layer was pre-baked at 85°C for 3 min and then exposed to UV light (I-line of Hg arc lamp, 365 nm) through a chromium-quartz mask at a dose of 90-100 mJ/cm^2. The pattern was then obtained by development in AR 300-26 for 45 s, followed by rinsing in deionized water and drying by blowing nitrogen. The device was then closed with a second CaF_2 window 1 mm thick through a thermo-mechanical process which exploits the residual solvent trapped inside the resist as a consequence of a milder resist pre-bake with respect to conventional parameters. The two windows were brought in intimate contact by a hydraulic press whose plates were heated to 65°C. At this temperature, a pressure of 5 bar was applied for 5 minutes, then the system was cooled down to room temperature; finally, the hereby sealed device was released. On the bottom window, 4 holes have been drilled before the lithographic process, in order to allow the connection with the external world assured by 4 L-shaped channels produced into the plastic holder shown in Figure 2c. The device thermalization is done by inserting the assembly in a cylindrical metallic heater. The temperature is set to 37 °C by a PID electronic controller which reads the temperature measured with a K thermocouple fixed onto the heater.

2.2. Data acquisition and analysis

Just before measurements, U937 monocytes were removed from the culture medium, that was exchanged with physiological solutions, and the cell suspension was injected inside the fluidic device using a syringe pump. IR microspectra of single living U937 monocytes have been acquired at 10x10 μm at SISSI beamline averaging 256 scans. A buffer spectrum was also acquired with the same parameters in a point close to the cell. Air background was collected within the appropriate device compartment. The "pure" cellular spectrum was then retrieved from the raw cell one by subtracting the buffer spectrum applying a scaling factor of 1 (Method 1) or adjusting it in order to match the water

Journal of Physics: Conference Series **359** (2012) 012007 doi:10.1088/1742-6596/359/1/012007

combination band of the two centered at ~2125 cm^{-1} (Method 2). In respect to Method 2, a MATLAB algorithm was written that minimizes the residual area between raw cell and buffer microspectra in the 2500-1850 cm^{-1} spectral region.

a

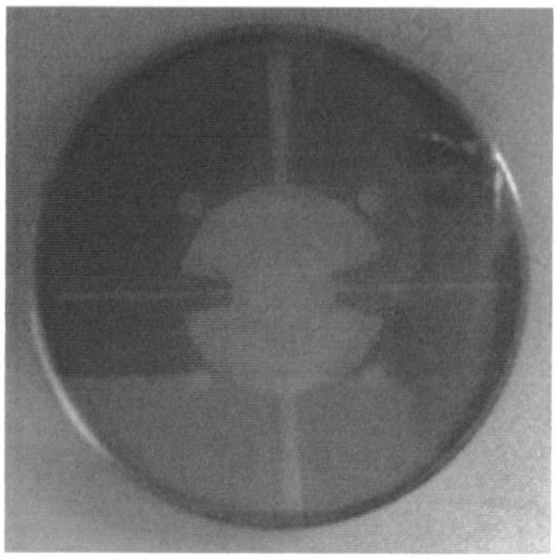

Figure 2a. Actual device obtained by lithographing XARP resist on CaF_2 window. It is made by two concentric chambers separated by a porous septum.

b

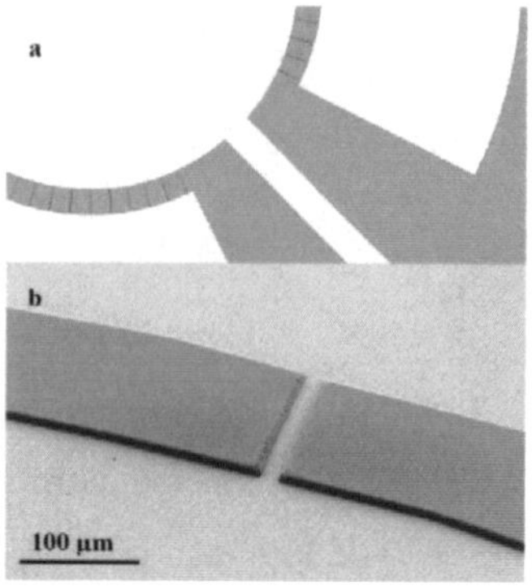

Figure 2b. Details of the device. In the upper figure the CAD lay-out of the porous septum with the inner chamber inlet; the lower is a SEM micrograph of the actual septum.

c

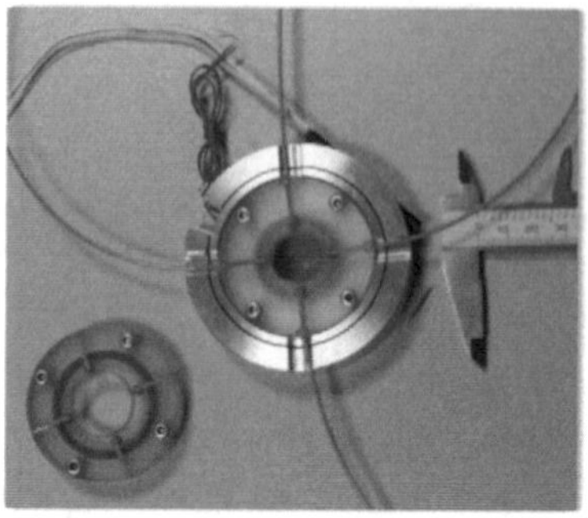

Figure 2c. Assembly of the micro-fluidic chip inside the external metallic heater and the fluidic connection system.

3. Results and discussion

3.1. Microfabrication of fully-sealed microfluidic devices

Biological MicroElectroMechanical Systems (Bio-MEMS) are analytical platforms where biological matter, from single molecules to entire living organisms such as bacteria, is manipulated in order to measure its response to different kind of stimuli [9, 10]. They are devices that comprise a wide variety of miniaturized structures, including moving parts (cantilevers, valves, and switches), static elements (channels, wells and porous sects), chemically functionalized surfaces (for promoting cell adhesion, as an example) and electrical components (resistors, T- and pH-probes). Bio-MEMS have revolutionized biological and medical science by reducing the reagent volumes for analysis and related costs, by shortening the time for analysis due to the higher effective concentrations reached within the devices, by increasing the throughput of many investigation techniques thanks to the parallelization of the detection and for many other reasons. However, such novelty did not impact IRMS and the reason has to be ascribed to the material requirements that have to be satisfied for producing IR-suitable bio-MEMS. Among the materials transparent to both Mid IR and visible light, the ones suitable for being microfabricated should be also biocompatible, economic and easily workable; however, none of the most common ones fulfil all the requirements. Fused silica, that satisfies all the Bio-MEMS requirements, has very poor transmission properties below 2500 cm^{-1}, while the good biological, optical and fabrication properties of diamond are paid at great cost. ZnSe is toxic for cellular samples [11] and also poorly transparent to visible light, while BaF_2 is quite soluble in water and therefore not suitable for our purposes. Calcium fluoride is matching almost all the requirements, except for the manufacturability. Fragile and with low thermal conductivity, CaF_2

cannot be easily "subdued" to microfabrication rules and the work done by our group was to optimize all the fabrication steps for produce a fully-sealed microfluidic device suitable for IR analysis, from the resist choice to the sealing step, included the photolithographic layout definition. It is beyond the aim of this paper to detail the optimization of fabrication protocols, briefly described in the experimental session and already detailed elsewhere [12] while we would stress on the advantages offered by microfluidic devices for IRMS analysis of living cells. The major improvement brought by the micro-approach is the design flexibility intrinsic in the method. Just as an example, the layout of one of the fluidic devices employed for the experiments presented in subsection 3.2 is reported in Figure 2a. It is made by two concentric chambers, separated by a porous septum, both connected with the external world via inlet-outlet macro-channels (see Figure 2b for details). The inner chamber accommodates the cells while the second is a solution reservoir where fresh physiological solution or chemical agents can be injected in, that will diffuse towards the cells while cell metabolites will diffuse out by osmosis. Syringe pumps are used for supplying both liquid and cell suspension, maintained at a constant temperature by a heater made by a resistor tape embedded into a stainless steel holder (see Figure 2c). The fluidic device accommodate also 4 isolated measurement points where air spectra can be collected without disassembling the device, as required by demountable liquid cells. Moreover, in respect to demountable liquid cell, the control of the device height is much more accurate and reproducible, both aspects that impact on the accuracy of water subtraction, as will be detailed later on.

For increasing the device throughput in the nucleic acid – carbohydrate spectral region, we implemented a sputtering process for deposit very thin silicon layers (10-20 nm) on top of BaF_2 optical windows, in order to avoid its dissolution in water with consequent cytotoxic effects on cellular sample and decreased IR-Vis transparency. Moreover, the silicon interface has two other major advantages:

1- It allows to employ more conventional resist, such as SU-8 or PMMA, as well as better established fabrication protocols
2- It allows exploiting the silane chemistry for the functionalization of the surface directly in contact with cells, for better mimicking the physiological conditions.

3.2. SR-IRMS of single living U937 monocytes

By using CaF_2 microfluidic devices, we performed several experiments in order to monitor the biochemical response of U937 monocytes to different kind of stimuli, both mechanical and chemical. These experiments will be summarized in this subsection for highlighting both the advantages and the disadvantages of single-cell IRMS with SR under physiological conditions.

The crucial step in the pre-processing of IR microspectra of living cells is the compensation for water features, in order to disclose the cellular spectral details, and in particular the protein bands Amide I and Amide II. For achieving this goal, the water contribution has to be subtracted, following standardized criteria. On the problems related to the reproducibility of spectral subtraction of water for the determination of protein secondary structure from dilute protein solutions, many authors have already reported [13, 14]. The faster way, named Method 1 hereafter, is to ratio the aqueous protein single-beam spectrum against the blank buffer reference. This method bases on two major assumptions: *i*- there are not significant differences in pathlength between recording the solvent and the protein single channel; *ii*- the water spectrum contributes for the same amount to the protein mixture and the buffer. By choosing a buffer point close to the acquired cell, the former criterion is quite well satisfied while the latter assumption can be considered almost always satisfied only for dilute protein solutions. This is obviously not the case of cellular spectra: rationing the cell single channel over the buffer single channel produces a microspectrum clearly over-subtracted, as the one shown in figure 3a, where Amide I and Amide II have almost the same intensity. A second route (Method 2 hereafter) can be followed for the pre-processing of living cell microspectra, similarly to the one most commonly employed also for dilute protein solutions: recording a reference spectrum of the air inside the device before the single channel either of the cell or of the buffer, and to ratio each single-beam spectrum against the most recent air-reference. At this point the "pure" cellular spectrum can be retrieve from the raw cell one by subtracting the buffer contribution. In this approach, the major source of uncertainty lies in the scaling factor to be applied. One possible criterion of choice, reported also by other author [15, 16], is the flatting of the spectral region between 1800 and 2500 cm^{-1}. This region is almost free from cellular spectral features and it is dominated by the water band centered at ~2125 cm^{-1}, combination of bending band (ν_2) and librational modes of water. The dependence on hydrogen bonding of water librational modes makes them sensitive to changes in liquid and solution structuring, and this represents the real insurmountable limit of this approach, at least at the actual stage of knowledge. Intracellular water structure and organization is quite complex and still debated [17] but for sure it greatly departs from the dilute solution state of the buffer medium. However, there are not other water reference bands that can be used, since all of them fall in spectral regions non cellular free. This implies that a countless degree of uncertainty is intrinsic in the technique and major efforts should be done for avoiding other source of errors. In this respect, microfabrication is fundamental in achieving more controllable conditions. Indeed, the possibility to use microfluidic devices containing dedicated wells for the collection of the air background is quite helpful, since an individual cell can be monitored for long time, even days, re-collecting background every wished time, without the need to open the device as in the case of demountable cells. The repeatability of Method 2 can also be improved in respect to the current approaches where the choice of the best scaling coefficient is carried out by using interactive graphic programs, therefore being user-biased. In this regard, we developed a MATLAB routine for minimizing the residual area between the combination band of cell and buffer microspectra by varying the scaling coefficient between 0.8-1. The developed automated algorithm was used for the pre-processing of data presented in the following.

It has to be highlight that the difficulties associated to data pre-processing for circumventing the water absorption barrier are partly compensated by the fact that the closer matching of the refractive indexes of cells [18, 19] and water [20, 21] with respect to air greatly suppress the Mie-type scattering phenomena that deeply affect microspectra of single fixed cells [22]. This can be easily appreciated by comparing Figure 1b with Figures 3a and 3b, that show the microspectra of the same single U937 cell obtained by applying Method 1 and 2 respectively.

a

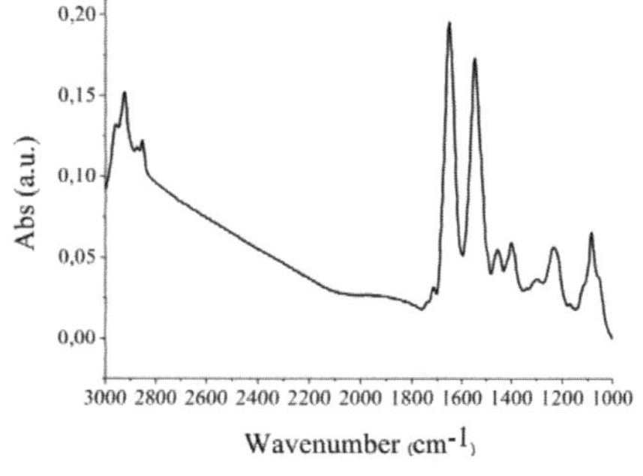

Figure 3a. SR-IR microspectrum of an individual living U937 monocyte obtained by applying the Method 1

b

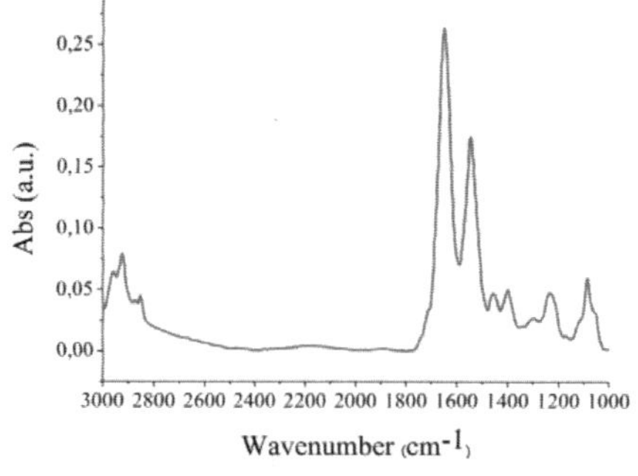

Figure 3b. SR-IR microspectrum of an individual living U937 monocyte obtained by applying the Method 2. Scaling factor determined in 0.924 via MATLAB routine

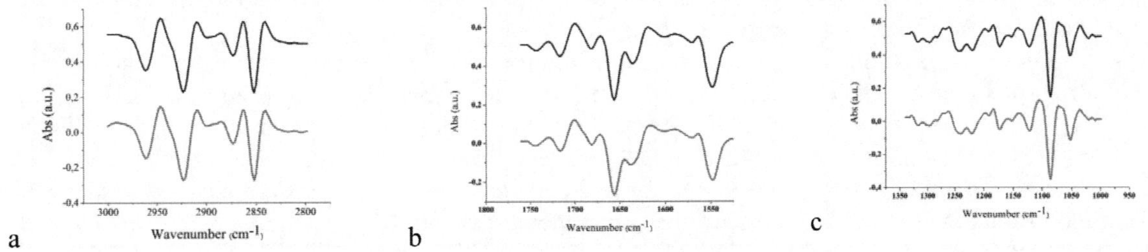

Figure 4a, b, c. Second derivatives of U937 microspectra (Savitzky-Golay algorithm, 13 smoothing point) obtained with Method 1 (black line) and 2 (red line) in the spectral regions a: 3000-2800 cm^{-1}, b: 1760-1480 cm^{-1} and c: 1350-1000 cm^{-1}.

Methods 1 and 2 are almost equivalent from a qualitative point of view, as evaluated by second derivative analysis (see figures 4a-c). No appreciable shifts in band position can be seen within the entire spectral range, with the exception of the broadening toward ~1645 cm^{-1} of the components of Amide I band in Method 2, accounting for the lower water subtraction than Method 1. On the contrary, many differences can be appreciated between the two approaches from a semi-quantitative point of view. As expected, the Method 1 is underestimating the integral intensity of both Amide I band (5.72 a.u. versus 11.87 a.u.) and, at a less extent, Amide II (5.97 a.u. versus 6.85 a.u.). This reflects in the lowering of Amide I over Amide II (AmI/AmII) ratio for the Method 1 (0.96) in respect to Method 2 (1,73). However, the AmI/AmII ratio retrieved with Method 2 is higher than ~1.5, commonly found in dried samples. This could account for structural water molecules, integral part of protein structure and fundamental for their biological activity [23, 24]. While the integral of DNA asymmetric band is equivalent by using the two methods, both lipid content and composition are method-sensitive: the total lipid content is overestimated (3.523 a.u. versus 2.906 a.u), while the methylene on methyl symmetric stretching ratio is underestimated (1.167 a.u. versus 1.285 a.u.). Similar results have been obtained for all the cellular samples investigated (data not shown) and compared with that coming from single cells fixed in formalin (data not shown), that is known to preserve both lipid content and composition. We found a better agreement between the formalin dataset and the one obtained with Method 2, which therefore we decided to systematically apply for the data pre-processing of living cell spectra.

Even if the application of the Method 2 is more time consuming, it offers some practical advantages beside the one previously highlighted. First of all, the acquired dataset can be always brought back to one obtained with Method 1, rationing the cell single channel over the buffer single channel, used as background. Secondly, from the buffer spectrum rationed over air background it is easier to check for the buffer cleanness. As a matter of fact, cells both excrete metabolites and chemical substances, in response to either external stimuli or naturally occurring events, for example the Extra Cellular Matrix (ECM), composed by fibrous proteins and glycosaminoglycans in the majority. The local accumulation of such cellular products can be quite high if the buffer medium is not properly exchanged, that is with a velocity sufficient for keeping cells healthy by providing fresh nutrients but lower enough for the completion of cellular events that need extracellular mediators, such as cell adhesion or duplication. Demountable liquid cells cannot be fluxed without the risk to push the investigated cell out from the field of view while extremely low fluxes do not guarantee an appropriate medium exchange. All these drawbacks can be solved by using a liquid cell with the design shown in Figure 2a, where a porous septum is dividing the cellular compartment from the medium reservoir that is therefore exchanged by slow diffusion.

It is evident that the exploitation of the advantages offered by microfabrication in the field of IRMS of living single cell with SR can open new and existing opportunities in monitoring the cell biochemistry under physiological conditions. We already demonstrated the possibility of following in real time the biochemical alterations undergone by living cells subjected to both chemical and mechanical stimuli. By stimulating U937 monocytes with fMLP, a synthetic peptide that mimics the activity of bacterially derived peptides with formylated N-terminal methionine groups, able to promote the monocyte extravasation for reaching the inflammation site, we monitored the synthesis of new adhesion proteins and the cell cytoskeleton rearrangement responsible for cell extravasation [25]. Similarly we monitored the response to U937 to mechanical deformation, confining them in fluidic devices of different thickness in order to achieve undeformed, deformed and heavily deformed conditions. We demonstrated that deeply deformed cells have a cellular biochemistry quite different from undeformed and slightly deformed ones, evidencing that the limit in fluidic device pathlength, imposed by the saturation of water bending band at ~9 microns, is in turn limiting the type of cells can be sampled to the ones that have an height lower that the pathlength constrain [26]. However, even if most of the circulating cells are usually larger than 8-9 microns, the majority of the adherent cells are not.

Finally, is has to be highlighted that a living cell never sleep. Therefore cell mapping, a quite long measurement, is not reliable since the risk is to start measuring an object and to finish with a completely different one. Therefore, fast SR chemical imaging is presently the more convenient approach for the fast collection of whole-cell cartograms with micrometric spatial resolution [27].

4. Conclusions

The extension of microfabrication concepts for producing fluidic devices where analyse living cells can promote the upgrade of SR-IRMS toward a mature label-free single cell-based assay. The design flexibility, one of the major advantages of the technique, allows performing innovative experiments never done before. We opened the road for the fabrication of CaF_2 devices, and we are testing new devices in BaF_2 silicon coated, that allow an easier access to nucleic acids bands and to tune the device surface properties by silane chemistry. However, even if we believe we did very big steps in optimizing fluidic for IRMS, some spectroscopic aspects have not be clarified yet and need further in-depth examinations. First of all, a standardized protocol has to be developed for the subtraction of water content from raw cell microspectra, in order to disclose Amide I band and give more reliable information also on cellular lipids. Microfabrication can help in pursuing this goal, but water is still imposing heavy limitations in cellular analysis, especially on cell dimensions. Finally, the advantage of monitor living cells in real time is imposing measurement times shorter than the time for completion of the phenomenon under investigation. In particular, fast imaging techniques are needed for pursuing living single cell analysis at sub-cellular spatial resolution, since a cell is always evolving.

5. References

[1] Holman H-Y N, Martin M C and McKinney W R 2003 Synchrotron-Based FTIR Spectromicroscopy: Cytotoxicity and Heating Considerations *Journal of Biological Physics* **29** 275-86
[2] Carr G L 2001 Resolution limits for infrared microspectroscopy explored with synchrotron radiation *Rev. Sci. Instrum.* **72** 2-7
[3] Dumas P 2003 The use of synchrotron infrared microspectroscopy in biological and biomedical investigations *Vibrational Spectroscopy* **32** 3-21
[4] Tobin M J, Puskar L, Barber R L, Harvey E C, Heraud P, Wood B R, Bambery K R, Dillon C T and Munro K L 2010 Vibrational Spectroscopy FTIR spectroscopy of single live cells in aqueous media by synchrotron IR microscopy using microfabricated sample holders *Vibrational Spectroscopy* **53** 34-8

[5] Nasse M J, Ratti S, Giordano M and Hirschmugl C J 2009 spectroscopic techniques Demountable Liquid / Flow Cell for in Vivo Infrared Microspectroscopy of Biological Specimens *Applied Spectroscopy* **63** 1181-6

[6] Marcsisin E J, Uttero C M, Miljkovic M and Diem M 2010 Infrared microspectroscopy of live cells in aqueous media *Analyst* **135** 3227-32

[7] Naumann D FT-IR and FT-NIR Raman Spectroscopy in Biomedical Research

[8] Lupi S, Nucara A, Perucchi A, Calvani P, Ortolani M, Quaroni L and Kiskinova M 2007 Performance of SISSI , the infrared beamline of the ELETTRA storage ring *Journal of the Optical Society of America B* **24** 959-64

[9] Spatz J P 2005 Bio-MEMS: building up micromuscles. *Nature materials* **4** 115-6

[10] Yoo S K, Lee J H, Yun S-S, Gu M B and Lee J H 2007 Fabrication of a bio-MEMS based cell-chip for toxicity monitoring. *Biosensors & bioelectronics* **22** 1586-92

[11] Kim J, Park Y, Yoon T H, Yoon C S and Choi K 2010 Phototoxicity of CdSe/ZnSe quantum dots with surface coatings of 3-mercaptopropionic acid or tri-n-octylphosphine oxide/gum arabic in Daphnia magna under environmentally relevant UV-B light *Aquat Toxicol* **97** 116-24

[12] Birarda G, Grenci G, Businaro L, Marmiroli B, Pacor S and Vaccari L 2010 Fabrication of a microfluidic platform for investigating dynamic biochemical processes in living samples by FTIR microspectroscopy *Microelectronic Engineering* **87** 806-9

[13] Barrett J J, Siebert D R, Reddy K V, Neusser H J, Schlag E W, Schenck P K, Leroi G E, Myerholtz C A, Enke C G, Nugent W H, Novak M, Cone R L, Suzuki T, Kato H, Konishi N, Kasuya T, Bennett J A, Birge R R, Rohrer F, Zetzseh C, Focus L, Moore D S, Keller R A, Engleman R, Alamos L, Steudel R, Wharton L and Levy D H 1989 On the Spectral Subtraction of Water from the FT-IR Spectra of Aqueous Solutions of Proteins *Focus* **43**

[14] Dong a, Huang P and Caughey W S 1990 Protein secondary structures in water from second-derivative amide I infrared spectra. *Biochemistry* **29** 3303-8

[15] Rahmelow K I M and Hu W 1997 Infrared Spectroscopy in Aqueous Solution: Difficulties and Accuracy of Water Subtraction *Applied Spectroscopy* **51** 160-70

[16] Venyaminov S Y and Prendergast F G 1997 Water (H 2 O and D 2 O) Molar Absorptivity in the 1000 – 4000 cm 01 Range and Quantitative Infrared Spectroscopy of Aqueous Solutions 1 **245** 234-45

[17] Jasnin M, Moulin M, Haertlein M, Zaccai G and Tehei M 2008 Down to atomic-scale intracellular water dynamics *EMBO Rep* **9** 543-7

[18] Romeo M and Diem M 2005 Correction of dispersive line shape artifact observed in diffuse reflection infrared spectroscopy and absorption/reflection (transflection) infrared micro-spectroscopy *Vibrational Spectroscopy* **38** 129-32

[19] Mohlenhoff B, Romeo M, Diem M and Wood B R 2005 Mie-Type Scattering and Non-Beer-Lambert Absorption Behavior of Human Cells in Infrared Microspectroscopy *Biophysical Journal* **88** 3635-40

[20] Schiebener P, Straub J, Sengers L J M H and Gallagher J S 1990 Refractive index of water and steam as function of wavelength, temperature and density *Journal of Physical and Chemical Reference Data* **19**

[21] QUERRY M R, CURNUTTE B and WILLIAMS D 1969 Refractive Index of Water in the Infrared *JOSA B* **59** 1299-304

[22] Bassan P, Byrne H J, Bonnier F, Lee J, Dumas P and Gardner P 2009 Resonant Mie scattering in infrared spectroscopy of biological materials – understanding the 'dispersion artefact' *The Analyst* **134** 1586

[23] Tarek M and Tobias D J 2002 Role of Protein-Water Hydrogen Bond Dynamics in the Protein Dynamical Transition *Physical Review Letters* **88** 138101

[24] Richards F M 1977 Areas, Volumes, Packing, and Protein Structure *Annual Review of Biophysics and Bioengineering* **6** 151-76

[25] Birarda G, Grenci G and Vaccari L 2010 *Microscopy: Science, Technology, Applications and Education,* ed A Méndez-Vilas and J D Álvarez: Publisher: Formatex Research Center) pp 422-32

[26] Birarda G, Grenci G, Businaro L, Marmiroli B, Pacor S, Piccirilli F and Vaccari L 2010 Infrared microspectroscopy of biochemical response of living cells in microfabricated devices *Vibrational Spectroscopy* **53** 6-11

[27] Nasse M J, Walsh M J, Mattson E C, Reininger R, Kajdacsy-Balla A, Macias V, Bhargava R and Hirschmugl C J 2011 High-resolution Fourier-transform infrared chemical imaging with multiple synchrotron beams *Nat Meth* **8** 413-6

Journal of Physics: Conference Series **359** (2012) 012008 doi:10.1088/1742-6596/359/1/012008

Study of Barley Grain Molecular Structure for Ruminants Using DRIFT, FTIR-ATR and Synchrotron Radiation Infrared Microspectroscopy (SR-IMS): A Review

Peiqiang Yu [1]
Ministry of Agriculture Strategic Research Chair
College of Agriculture and Bioresources, University of Saskatchewan
51 Campus Drive, Saskatoon, Canada, S7N 5A8, Canada

E-mail: peiqiang.yu@usask.ca

Abstract. Barley inherent structures are highly associated with nutrient utilization and availability in both humans and animals. Barley has different degradation kinetics compared with other cereal grains. It has a relatively higher degradation rate and extent, which often cause digestive disorder in the rumen. Therefore understanding barley inherent structure at cellular and molecular levels and processing-induced structure changes is important, because we can manipulate barley inherent structures and digestive behaviors. Several molecular spectroscopy techniques can be used to detect barley inherent structures at cellular and molecular levels. This article reviews several applications of the IR molecular spectral bioanalytical techniques - DRIFT, FT/IR-ATR and SR-IMS for barley chemistry, molecular structure and molecular nutrition research

1. Introduction

Internal molecular structures of barley affect nutrient availability and fermentation characteristics of barley. For example, Valier and Harrington barley proteins have similar chemical composition but different degradation or fermentation behaviors [1-3]. Therefore, understanding the barley inherent structure at cellular and molecular levels and the processing-induced structure changes is important, because we can manipulate barley inherent structures and digestive behaviors. However, traditional "wet" chemical analysis fails to reveal barley internal structure and to link its inherent structural information to chemical one [1,4], due to the destruction of the structure during the processing for chemical analysis. IR molecular spectroscopy techniques can be used to detect barley inherent structures at cellular and molecular levels. The objective of this article is to review the applications of several IR molecular spectral bioanalytical techniques - DRIFT, FT/IR-ATR and SR-IMS for barley chemistry, molecular structure and molecular nutrition research in my team.

2. Application A: Microprobing Molecular Spatial Distribution and Structural Architecture of Barley with SR-IMS Technique

Yu [5] reported that the advanced Synchrotron-Radiation Infrared MicroSpectroscopy (SR-IMS) is able to study cell or living cell biochemistry within cellular dimension [4]. The author used the SR-

[1] To whom any correspondence should be addressed.

Published under licence by IOP Publishing Ltd

Journal of Physics: Conference Series **359** (2012) 012008 doi:10.1088/1742-6596/359/1/012008

IMS imaging to microprobe molecular spatial distribution and cell architecture of cereal grain tissue in a comprehensive way. Chemical images were collected by SR-IMS at the beamline U2B at the National Synchrotron Light Source, Brookhaven National Laboratory (NSLS-BNL, U.S. Department of Energy, New York). The molecular images were systematically carried out from the outside to the inside of the seed tissue under various chemical functional groups and their ratios. The results show that unique cell architecture and the molecular spatial distribution, intensity in the seed tissue could be generated using the SR-IMS. Figure 1 is one example. This imaging technique and methodology provide a high potential that could be used by scientists to develop a specific cereal grain variety with targeted food and feed qualify; moreover, it can also be used to monitor the degree of grain maturity and grain damage, the fate of organic contaminants and the effects of chemical treatment on plant and grain seeds.

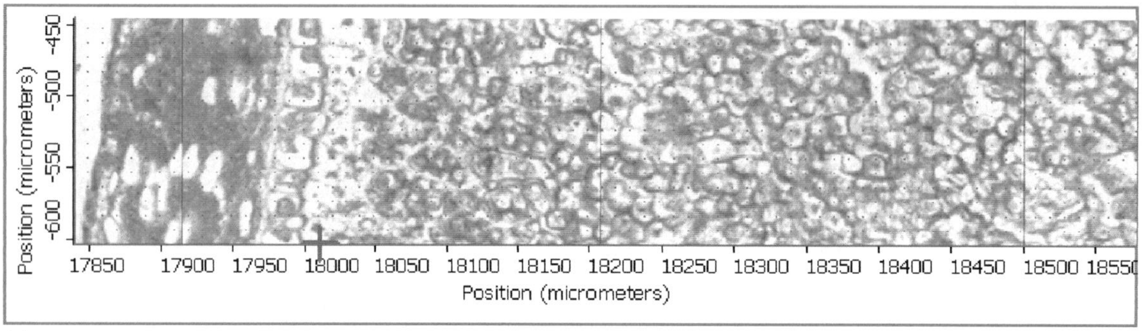

(a)

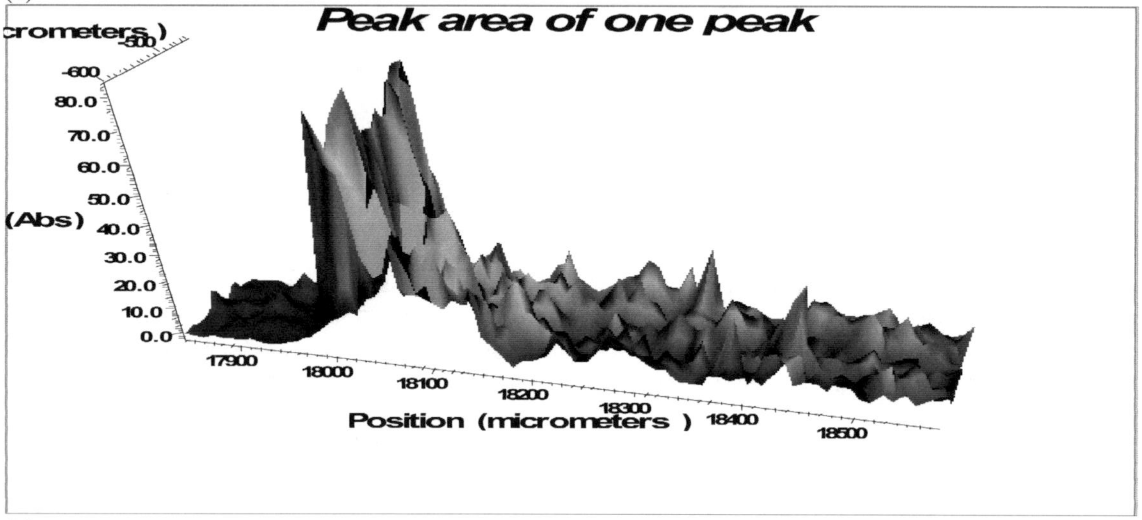

(b)

(c)

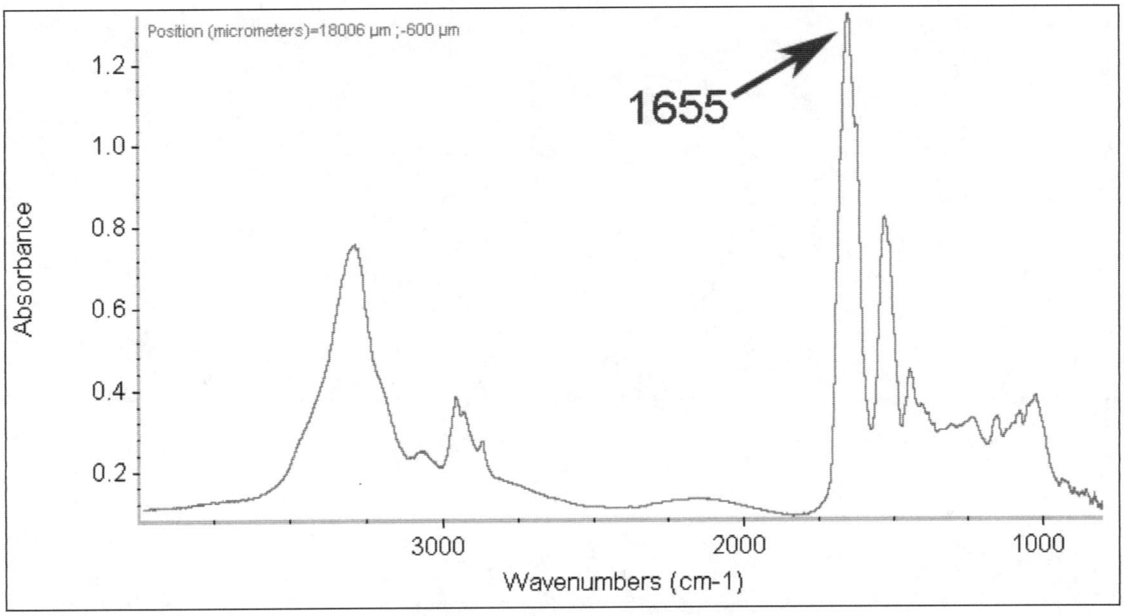

(d)

Figure 1. SR-IMS molecular functional group image from the pericarp (outside), to seed coat, aleurone layer and endosperm- a). Visible image; (b) 3D chemical image; (c) function group intensity ruler; and (d) Spectra corresponding to the pixel at the cross-hair in the visible image. Images were acquired with a pixel resolution of 10 ×10 µm at National Synchrotron Light Sources, Brookhaven National Laboratory (NSLS-BNL, U.S. Dept of Energy, New York) (Source: [5])

3. Application B: Molecular clustering, interrelationships and carbohydrate conformation in hull and seeds among barley cultivars

In this study [6], the authors applied molecular spectral analyses (univariate and multivariate) to Diffuse Reflectance Infrared Fourier Transform (DRIFT) spectra to study carbohydrate conformation features, molecular clustering and interrelationships in hull (Figure 2) and seed (Figure 3) among six barley cultivars, which had different degradation kinetics in rumen. The authors analyzed the spectral fingerprint regions in both hull and seeds together with Agglomerative Hierarchical Cluster (AHCA) and Principal Component (PCA) analyses. It was found that the carbohydrate molecular spectral analysis in the ca. 1185-800 cm^{-1} region together with the AHCA and PCA were able to show that the barley seed inherent structures exhibited distinctive differences among the barley varieties. CDC Helgason had discernible differences from AC Metcalfe, MeLeod, CDC Cowboy and CDC Dolly in carbohydrate conformation in the seed (Figure 4). Clear molecular clusters could be distinguished and identified by using AHCA analysis and the separate ellipses could be grouped in PCA analysis. But CDC Helgason had no detectable differences from CDC Trey in carbohydrate conformation. These carbohydrate conformation/structure differences could be partially explained since the varieties were different in digestive behaviors in animals. The authors concluded that the DRIFT technique combined with AHCA and PCA molecular analyses was able to reveal carbohydrate conformation features and identify carbohydrate molecular structure differences in both hull and seeds among the barley varieties. The molecular spectroscopy technique used in this study could also be used for other plant-based feed and food structure study [6].

Journal of Physics: Conference Series **359** (2012) 012008 doi:10.1088/1742-6596/359/1/012008

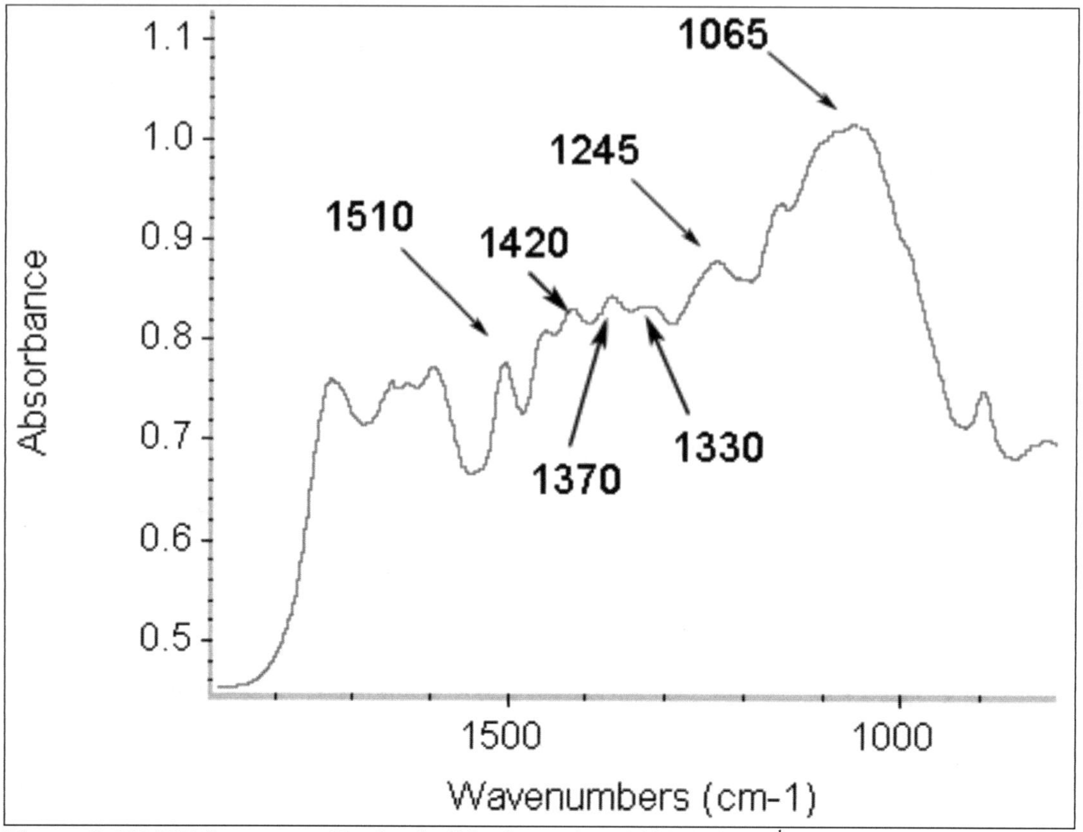

Figure 2. DRIFT Spectrum of barley hull in the region: 1800-800 cm^{-1} (Source: [6])

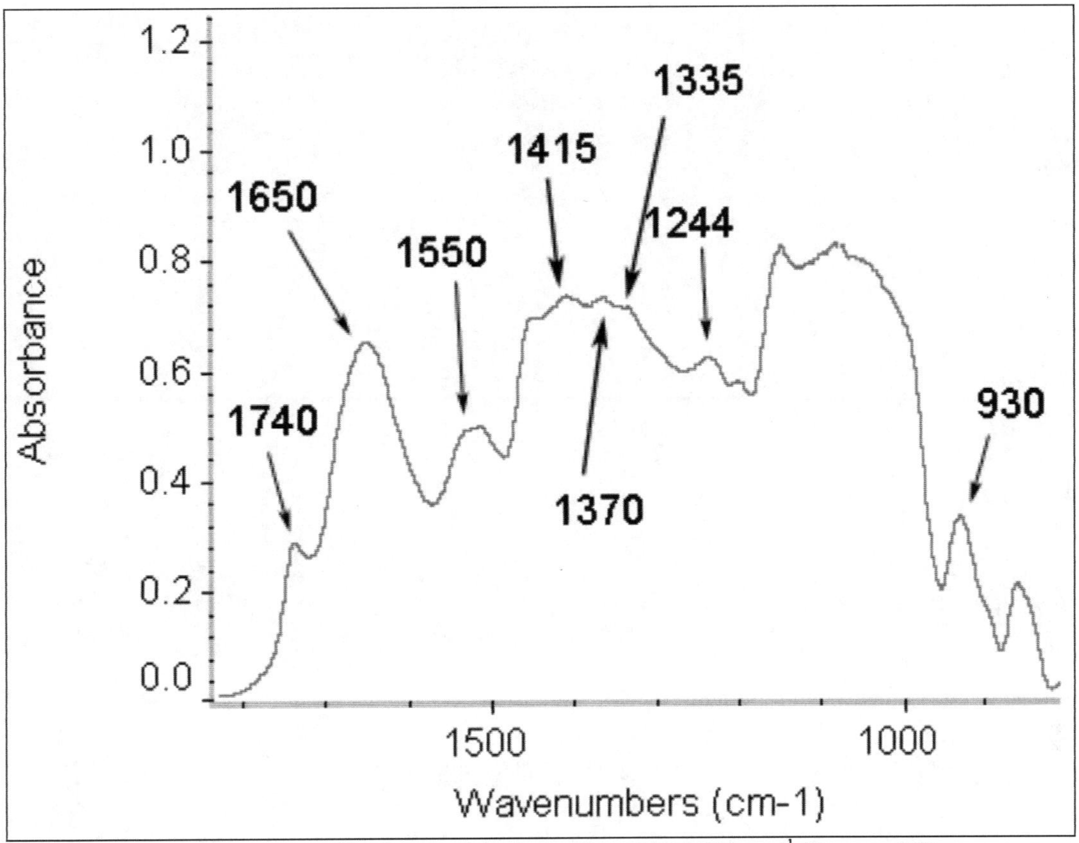

Figure 3. DRIFT Spectrum of barley seed in the region: 1800-800 cm^{-1} (Source: [6])

Journal of Physics: Conference Series **359** (2012) 012008 doi:10.1088/1742-6596/359/1/012008

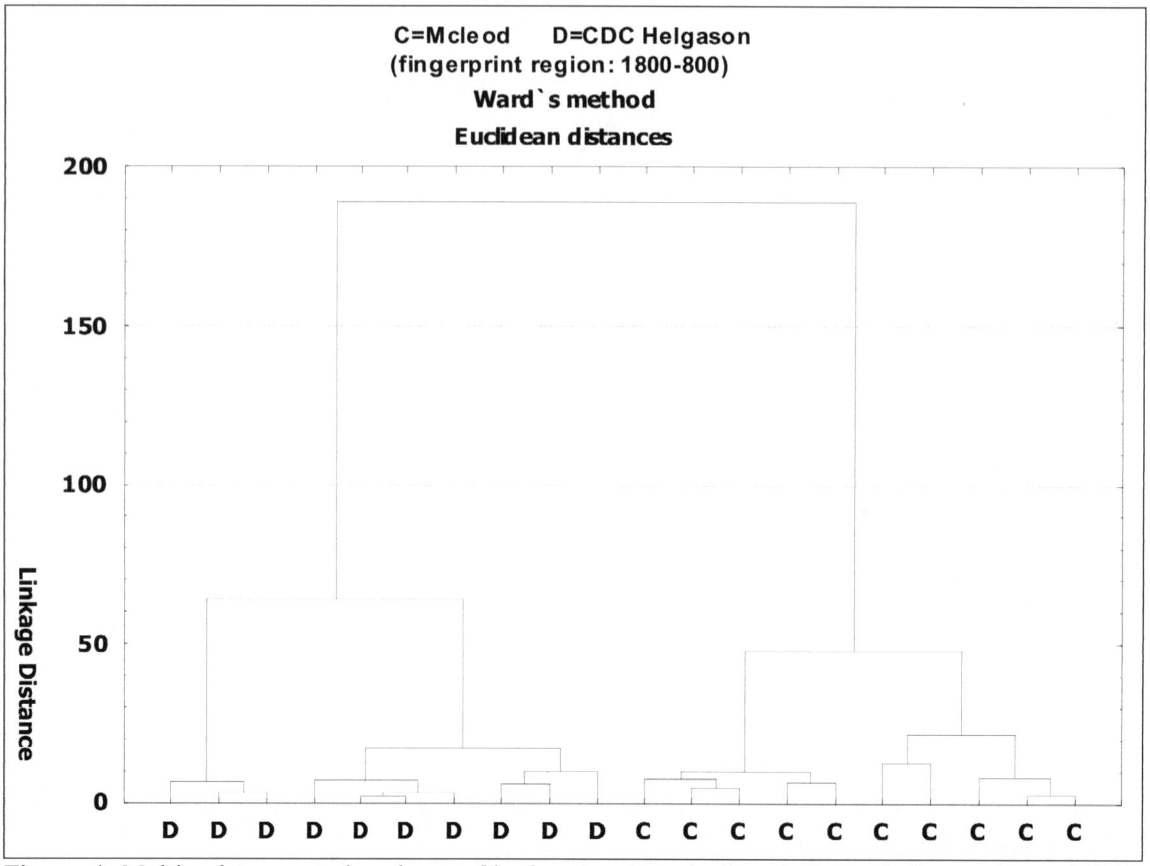

Figure 4. Multivariate spectral analyses of barley structures in the whole seed: Comparison of CDC Helgason (D) and MeLeod (C) using agglomerative hierarchical cluster (AHCA) (1) Select spectral region: finger print region: ~1800 to 800 cm^{-1}; (2) Euclidean distance; (3) Cluster method: Ward's algorithm. (Source: [6])

4. Application C: Using Molecular Spectroscopy with Uni- and Multi-variate Spectral Analysis Techniques to Detect Protein Molecular Structure Differences among Different Genotypes of Barley

In this study [7], the authors characterized protein molecular structure using DRIFT spectroscopy with univariate and multivariate molecular spectral analyses and identified the structure differences in both hull and seeds among six spring barley varieties. The molecular structure spectral analysis involved protein amide I and II regions ca. 1716-1485 cm^{-1} (attributed to protein amide I C=O and C-N stretching; amide II N-H bending and C-N stretching) [9,12] together with AHCA and PCA analyses. The authors found that the techniques were able to identify the spectral differences associated with the molecular structural differences among the barley varieties. The molecular spectral analyses in the region of ca. 1715-1485 cm-1 together with AHCA and PCA analyses were able to show that the molecular structures of the seeds exhibited distinctive differences among the barley varieties, while the hull did not. The authors also found that CDC Helgason is differentiated by AC Metcalfe, McLeod, and CDC Cowboy in both protein amide I and II. Figure 5 shows one example of PCA analysis [7]. The molecular spectral technique provides an approach for plant protein molecular structure and biopolymer conformation study [7,10-13].

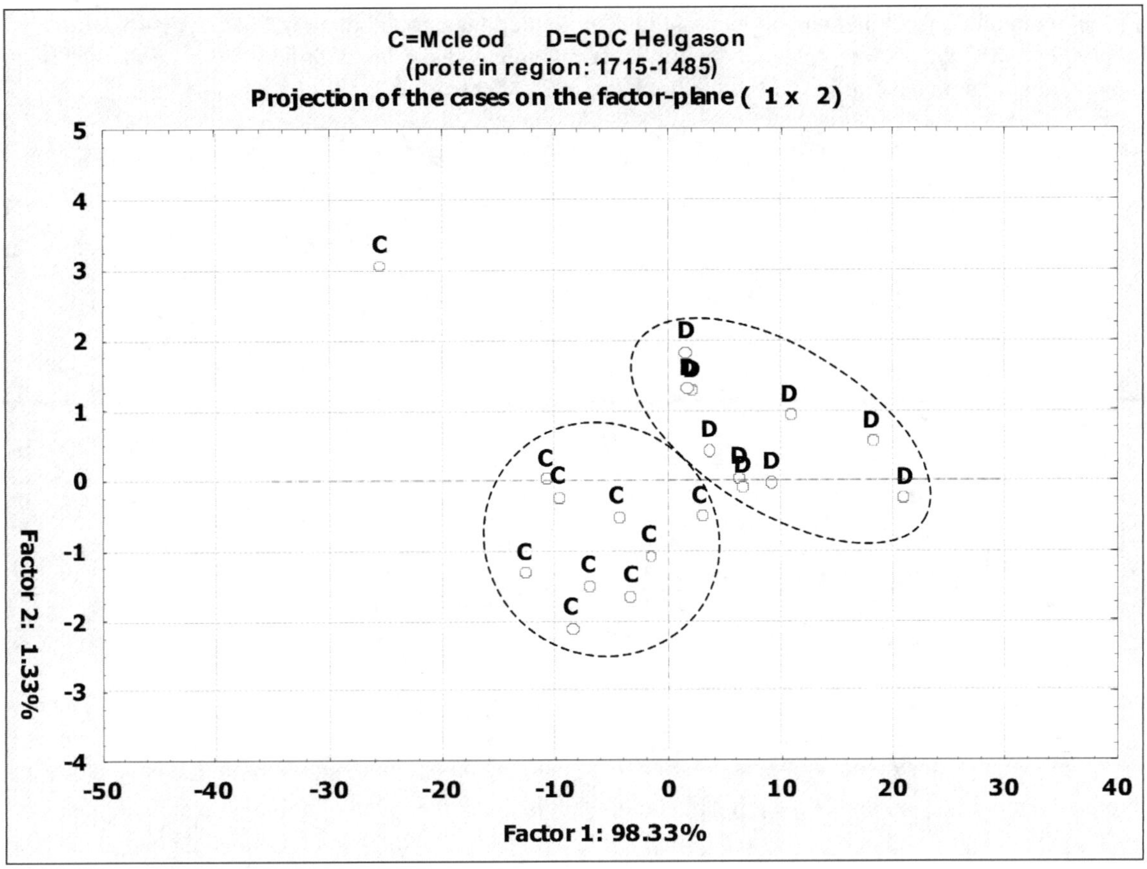

Figure 5. Multivariate spectral analyses of barley structures in the whole seed: Comparison of CDC Helgason (D) and McLeod (C): PC1 and PC2 explain 98.33 and 1.33% of the variances, respectively [principal component analysis: Scatter plots of the 1st principal components (PC1) vs. the 2nd principal components (PC2)] (Source: [7])

5. Application D: Characterization of the Microchemical Structure of Seed Endosperm within a Cellular Dimension Using Synchrotron-Based Infrared Microspectroscopy

In this study [8], the authors used SR-IMS to determine the microchemical structural features in seed endosperm tissue of six developed barley varieties and to study the relationship between molecular structural characteristics and degradation kinetics and nutrient availability in six genotypes of barley. The authors found that inherent microchemical structural differences in the endosperm among the six barley varieties were detected by the SR-based analytical technique. The SR-IMS spectral profiles differed ($P<0.05$) among the barley samples in terms of the peak ratio, peak area and height intensities of the amide I (ca. 1650 cm^{-1}) and amide II (ca. 1550 cm^{-1}) bands, cellulosic compounds (ca. 1240 cm^{-1}) and CHO component peaks (the 1st peak at the region ca. 1184-1132 cm^{-1}, the 2nd peak at ca. 1132-1066 cm^{-1}, the 3rd peak at ca. 1066-950 cm^{-1}) (Figure 6. With the SR-IMS technique, the structural characteristics of the cereal seeds were highlighted among different cultivars. The structural differences of barley seeds may be one reason for the various digestive behaviours and nutritive values for ruminants. The study [8] shows weak correlations between the functional groups' spectral data (peak area, height intensities and ratios) and rumen biodegradation kinetics (rate and extent of nutrient degradation). Weak correlations may indicate that limited variations of these six barley varieties might be not sufficient to interpret the relationship between spectroscopic information and nutrients value of barley grain, although significant differences in the biodegradation kinetics were observed. Finally the authors concluded that the studies demonstrated the potential of spatially resolved synchrotron based technology (SR-IMS) to reveal the structural and chemical make-up within cellular and subcellular dimensions without destruction of the inherent structure of cereal grain tissue.

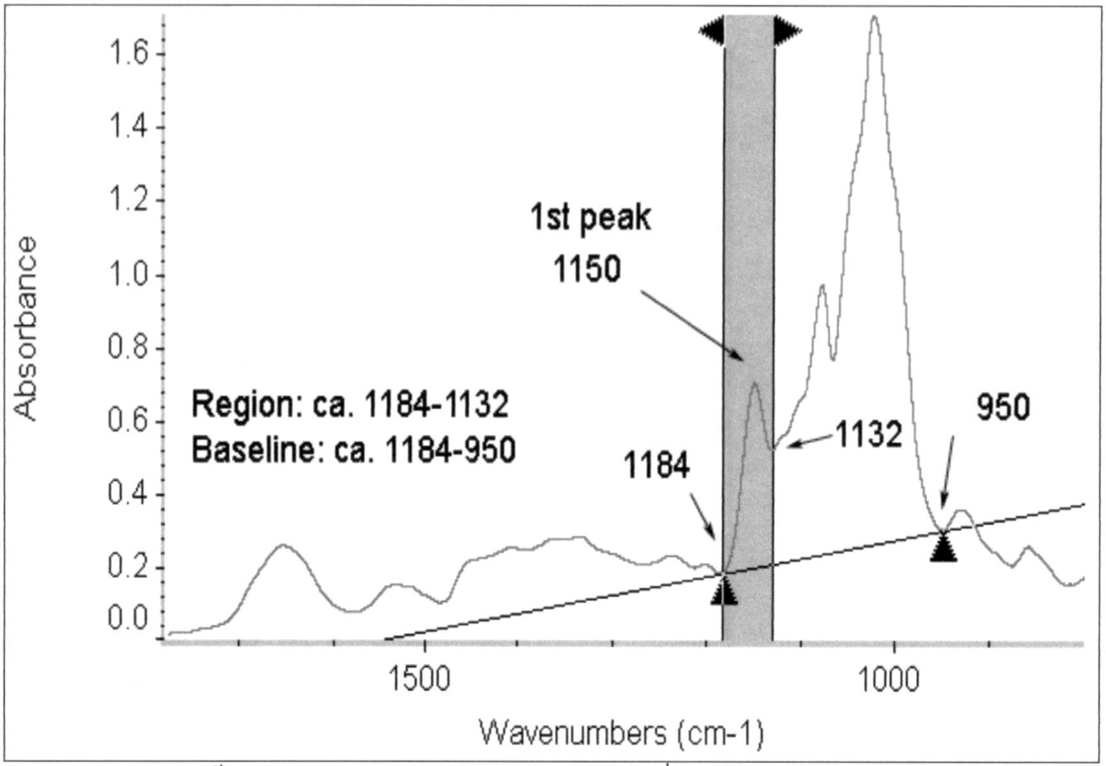

(a) Carbohydrate 1st peak height (Region: ca.1184-1132 cm^{-1})

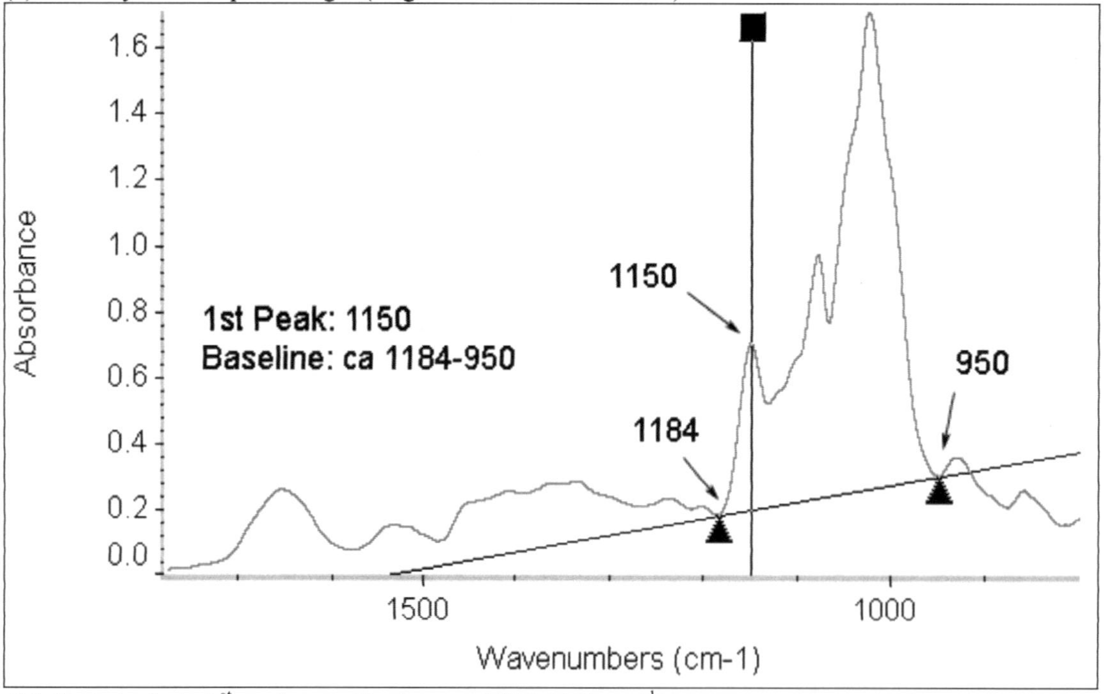

(b) Carbohydrate 1st peak area (Region: ca.1184-1132 cm^{-1})

Figure 6. Typical synchrotron-based FTIR spectrum of endosperm tissue within a cellular dimension: Carbohydrate 1st component peak area and height: ca. 1150 cm-1 (source, [8])

6. Conclusion:

In conclusion, the IR molecular spectral bioanalytical techniques- DRIFT, FT/IR-ATR and SR-IMS could be used for barley cell organization architecture, barley feed chemistry, molecular structure and molecular nutrition research. These techniques can be used to detect various processing-induced or related molecular structure changes and can be used to check processing efficiency.

References

[1] Yu P and Nuez-Ortín W G 2010. Relationship between Protein Molecular Structures and Metabolisable Proteins in Different Types of Dried Distillers Grains with Solubles: a novel approach. *British Journal of Nutrition (Cambridge University Press).* **104:** 1429–1437

[2] Damiran D, Yu P 2010. Chemical profile, rumen degradation kinetics and energy value of four hulless barley cultivars: comparison of the zero-amylose waxy, waxy, and high-amylose and normal starch cultivars. *Journal of Agricultural and Food Chemistry.* **58**:10553–10559

[3] Du L and Yu P 2011. Relationship of Physicochemical Characteristics and Hydrolysed Hydroxycinnamic Acids Profile (Ferulic acid, p-Coumaric Acid, Their Ratio) of CDC Barley Varieties and Nutrient Availability in Ruminants. *Journal of Cereal Science (Elsevier)* **53:** 178-187

[4] Budevska B O 2002. Vibrational Spectroscopy. Applications in Life, Pharmaceutical and Natural Sciences. John and Wiley and Sons, Ltd. Pp 3720-3732

[5] Yu P. 2011. Microprobing Molecular Spatial Distribution and Structural Architecture of Sorghum Seed Tissue (Sorghum Bicolor L.) with SR-IMS Technique. *Journal of Synchrotron Radiation.* In Press (VE5007).

[6] Liu N and Yu P 2011. Molecular Clustering, Interrelationships and Carbohydrate Conformation in Hull and Seeds among Barley Cultivars. *Journal of Cereal Science (UK).* **53**: 379-383

[7] Liu N and Yu P 2010a. Using DRIFT Molecular Spectroscopy with Uni- and Multi-variate Molecular Spectral Techniques to Detect Plant Protein Molecular Structure Difference among Different Genotypes of Barley. *Journal of Agricultural and Food Chemistry.* **58**: 6264–6269

[8] Liu N and Yu P 2010b. Characterize Microchemical Structure of Seed Endosperm within a Cellular Dimension among Six Barley Varieties with Distinct Degradation Kinetics, Using Ultraspatially Resolved Synchrotron-Based Infrared Microspectroscopy. *Journal of Agricultural and Food Chemistry.* **58**: 7801–7810

[9] Jackson M and Mantsch HH 1995. The use and misuse of FTIR spectroscopy in the determination of protein structure. *Biochemistry and Molecular Biology*. **30**: 95-120.

[10] Wetzel DL, Srivarin P and Finney JR 2003. Revealing protein infrared spectral detail in a heterogeneous matrix dominated by starch. *Vib Spectrosc* **31**: 109-114.

[11] Wang Q, Kretlow A, Beekes M, Naumann D and Miller L 2005. In situ characterization of prion protein structure and metal accumulation in scrapie-infected cells by synchrotron infrared and X-ray imaging. *Vib Spectrosc* **38**: 61-69.

[12] Marinkovic NS, Chance MR 2006. Synchrotron Infrared Microspectroscopy. Pages 2 671-708 in Encyclopedia of Molecular Cell Biology and Molecular Medicine, Wiley, Inc., Hoboken; 2006.

[13] Miller LM, Dumas P 2006. Chemical imaging of biological tissue with synchrotron infrared light. *Biochimica et Biophysica Acta*. **1757**: 846-857.

Journal of Physics: Conference Series **359** (2012) 012009 doi:10.1088/1742-6596/359/1/012009

Design of a dedicated beamline for THz coherent synchrotron radiation at UVSOR-III

Shin-ichi Kimura[1,2], Eiken Nakamura[1], Keiichiro Imura[1], Masahito Hosaka[3], Toshiharu Takahashi[4], and Masahiro Katoh[1,2]

[1] UVSOR Facility, Institute for Molecular Science, Okazaki 444-8585, Japan
[2] School of Physical Sciences, The Graduate University for Advanced Studies (SOKENDAI), Okazaki 444-8585, Japan
[3] Synchrotron Radiation Research Center, Nagoya University, Chikusa-ku, Nagoya 463-8603, Japan
[4] Research Reactor Institute, Kyoto University, Kumatori, Osaka 590-0494, Japan

E-mail: kimura@ims.ac.jp

Abstract. We report the design of a THz beamline for coherent synchrotron radiation (CSR) at the UVSOR-III very-low-emittance synchrotron radiation light source. The emitted THz-CSR is collected by a three-dimensional "magic mirror," which is a perfect collecting mirror for bending-magnet radiation with an acceptance angle of 288 mrad (H) × 80 mrad (V). A quasi-monochromatic THz-CSR with an average flux of 104 μW/0.1 % b.w. and a peak power of 120 nJ/pulse/0.1 % b.w. is expected at the beamline.

1. Introduction

Infrared (IR) and terahertz (THz) spectroscopies have now become conventional methods for the measurement of vibration modes and low-energy electronic structures. IR/THz synchrotron radiation (SR) is also used as a light source for spectroscopy and microscopy throughout the world. The reason is that IR/THz-SR is an ideal source, in which the electron beam emittance is smaller than the diffraction limit of light. So far, many advanced types of spectroscopy have been developed such as IR imaging, pressure-dependent and magnetic-field-dependent THz spectroscopy, and so on. [1] However, to realize further advances in spectroscopy, intense tunable light sources must be developed.

One candidate for an intense broadband THz source is coherent synchrotron radiation (CSR). [2] CSR is light with a longer wavelength than the length of the longitudinal fine structure of an electron bunch in an accelerator. The longitudinal fine structure then behaves like one particle with a large charge density. Since the length of the fine structure is similar to the wavelength of THz light, CSR appears in the THz region. The generation of CSR from storage ring has been investigated at several SR facilities, [3, 4, 5] especially using a laser-slicing method at Advanced Light Source, USA. [6] However, only few applications using CSR have so far been performed [7, 8], because of the instability.

At the UVSOR Facility, basic studies on the generation of CSR by the interaction between the electron beam in the storage ring and a pulse laser introduced from outside have been performed so far. Broadband CSR was produced using a bursting mode [9] and a laser slicing

Journal of Physics: Conference Series **359** (2012) 012009 doi:10.1088/1742-6596/359/1/012009

technique [10, 11]. Quasi-monochromatic CSR was also obtained due to the periodic modulation of electron bunches created by an amplitude-modulated laser pulse. [12, 13] The obtained quasi-monochromatic CSR had 10^4 times an intensity of "normal" (incoherent) SR. On the other hand, another type of coherent radiation, coherent harmonic generation (CHG), in the vacuum-ultraviolet (VUV) region was also studied using the same periodic modulation of electron bunches. [14, 15, 16]

Since 2008, a research project on the utilization of coherent light sources called Quantum Beam Technology Program has been under way, in which THz-CSR and VUV-CHG are being used for applications. The main purpose of the program is to develop novel types of spectroscopy using intense THz-CSR, such as THz pump-photoemission probe spectroscopy(TP^{3}S) to investigate the change of electronic structure after excitation of low energy electronic and vibrational states and quasiparticles. [17]

At the UVSOR Facility, the first SR was obtained from the old UVSOR-I ring in 1983. In 2003, the emittance of the light source was improved from 160 nm·rad to 27 nm·rad, and four undulators were installed (UVSOR-II). In 2012, the storage ring will be again improved to achieve an emittance of 15 nm·rad (UVSOR-III). The main targets of UVSOR-III are soft x-ray micro-spectroscopy and imaging, high-resolution angle-resolution photoemission spectroscopy in the VUV region, and application of coherent radiations.

In this paper, we report on the beamline dedicated to THz-CSR at UVSOR-III. The acceptance angle of the beamline is 288 mrad (H) × 80 mrad (V), which is larger than that of the previous IR/THz beamline BL6B, in UVSOR-II [215 mrad (H) × 90 mrad (V)]. [18] Although the horizontal acceptance angle of IRENI beamline at Synchrotron Radiation Center in Wisconsin, USA, is the largest [320(h)×25(v) mrad^2], [19] our new THz-CSR beamline has the largest solid angle. The emission intensity of THz-CSR is dependent on the emission angle of the bending magnet. To obtain intense light at around 1 THz, the design of the optics employs a three-dimensional "magic mirror" (3D-MM). As a result, the average flux and peak intensity at 1 THz or higher are expected to be 2×10^{18} ph/s/0.1 % b.w. and 120 nJ/pulse/0.1 % b.w., respectively.

2. Front end

Figure 1 shows the form factor of the THz emission from the electron bunch modulated by pulse laser as a function of the horizontal emission angle from the straight section. The form factor corresponds to the efficiency of the evolution of CSR. THz-CSR below 3 THz (~100 cm^{-1} ~12 meV) is generated from 10 degrees. The form factor of the 3 THz emission, however, is very small. An emission with a frequency lower than 2 THz is therefore mainly used in this beamline. Above 15 degrees, the intensity of THz-CSR below 2 THz (~67 cm^{-1} ~8 meV) rapidly increases. According to the geometrical conditions, the first mirror can be located at a maximum of 34 degrees. Because the maximum width of the first mirror was limited to 300 mm, the acceptance angle was set at 17.5–34 degrees (total 288 mrad) in the horizontal direction. The vertical angle was set at ±40 mrad to collect the widely expanded THz-CSR.

A top view of the THz beamline is shown in Fig. 2. The emitted THz light is collected by a 3D-MM (M0) of the same type as those already successfully installed at BL43IR in SPring-8 [20] and BL6B in UVSOR-II [18]. The 3D-MM was installed in the bending-magnet chamber and is controlled by a 5-axis pulse motor stage (x, z translation; θ_x, θ_y, θ_z rotation). To extract the emission at 34 degrees, the 3D-MM is located at the short distance of 55 mm from the electron orbit. The 3D-MM is made of aluminum and was fabricated by an NC length milling machine at the Equipment Development Center of Institute for Molecular Science. The surface roughness is less than ±1 μm, which is comparable to the visible wavelength but much smaller than the wavelength of THz light. Visible light is scattered by the surface of the 3D-MM, but THz light can be focused. To eliminate the heat load from the SR, we employed a copper rod

Journal of Physics: Conference Series **359** (2012) 012009 doi:10.1088/1742-6596/359/1/012009

(6 mm in diameter) with water cooling on the emission plane because the power from the SR is concentrated in the emission plane. The 3D-MM, which is located 880 mm from the center of the source, reflects the THz-CSR to the focal point (F0) on the outside of the bending-magnet chamber, as shown in Fig. 2 (1200 mm from M0). The THz-CSR collected by the 3D-MM

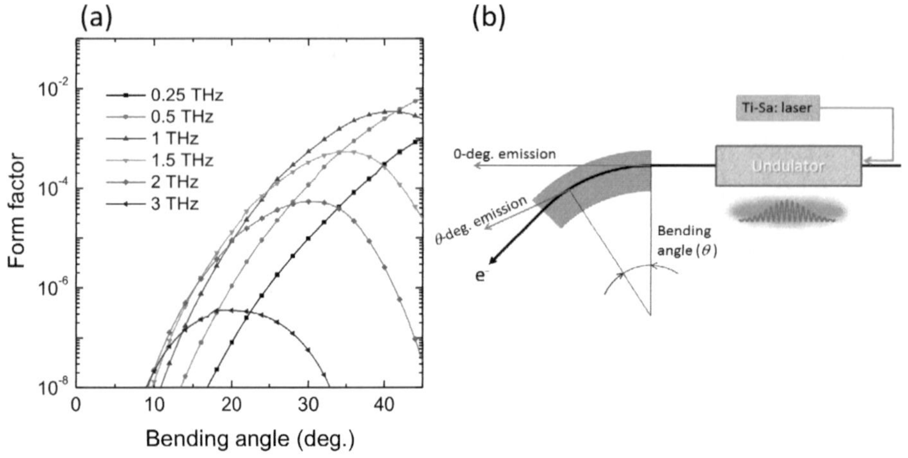

Figure 1. Frequency-dependent form factor of THz-CSR as a function of the horizontal angle of the bending magnet from the straight section (a) and the schematic layout of the emission angle of the bending-magnet radiation (b). The interaction between the electron bunch and pulse laser light occurs in the upstream undulator.

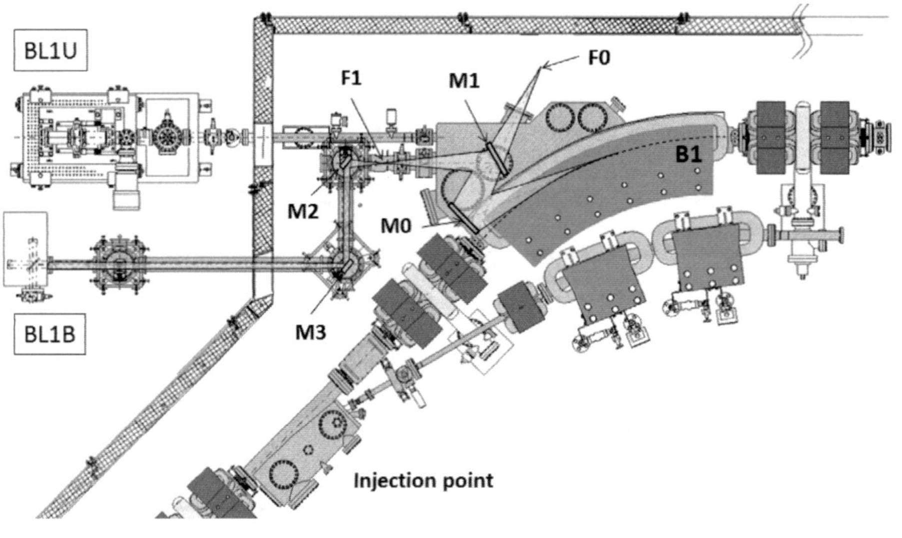

Figure 2. Schematic top view of the beam extraction part of the THz-CSR beamline, BL1B, at UVSOR-III. The three-dimensional magic mirror (3D-MM, M0) and a plane mirror (M1) are located in the bending-magnet chamber. A parabolic mirror (M2) is installed to form a parallel beam. The straight section (BL1U) is used for coherent harmonic generation (CHG) in the VUV region.

is intercepted by a plane mirror (M1) located in the bending-magnet chamber and guided to the beamline. The focal point (F1) is located on the outside of the bending-magnet chamber through a tapered z-cut quartz window (24 mm in diameter and 5 mm thickness at the center). After F1, the light is formed into a parallel beam by a parabolic mirror (M1) and guided to the outside of the radiation shielding wall.

A new vacuum chamber containing the bending magnet as well as the 3D-MM and M1 plane mirrors was installed in the spring of 2011. In the beamline, a Martin-Puplett type Fourier transform interferometer (FARIS-1, JASCO ltd.) will be installed to check the spectral features of the THz-CSR. Installation of all components of the THz beamline will be completed by December 2011.

3. Photon flux

The calculated photon flux spectrum of THz-CSR compared with "normal" SRs at BL1B and the old BL6B is shown in Fig. 3. Quasi-monochromatic CSR is obtained using the periodic modulation of electron bunches created by an amplitude-modulated laser. The bandwidth is about 0.1 % of the frequency. The frequency of the light can be scanned by changing the periodicity of the amplitude-modulated laser. The CSR spectrum intensity in Fig. 3 is the trace of the peak photon flux of the quasi-monochromatic CSR. The CSR spectrum has a peak at about 1 THz and the intensity below 3 THz is superior to that of normal SR. The expected average flux and peak power at 1 THz are about 2×10^{17} ph/s/0.1 % b.w.(=104 μW/0.1 % b.w.) and 120 nJ/pulse/0.1 % b.w., because the repetition rate of the pulse laser for modulation is 1 kHz.

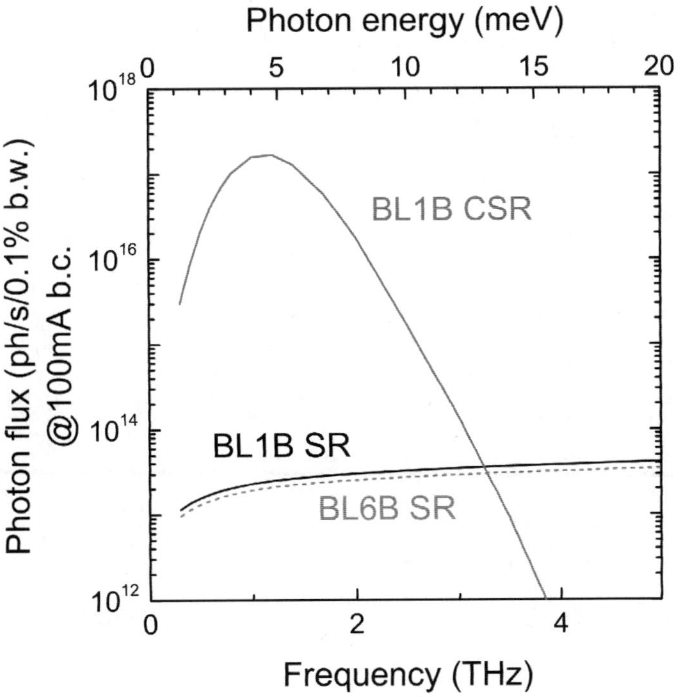

Figure 3. Calculated spectrum of the photon flux of THz-CSR at BL1B of UVSOR-III in comparison with the normal SRs emitted at BL1B and BL6B of UVSOR-III.

4. Conclusion

We have designed a beamline for coherent synchrotron radiation in the terahertz region (THz-CSR). An acceptance angle of 288 mrad(H)$\times$80 mrad(V) was set at a bending radius of 17.5–34 mrad due to the form factor of the CSR. The average flux and peak power at 1 THz or higher are expected to be about 104 μW/0.1 % b.w. and 120 nJ/pulse/0.1 % b.w., respectively.

Acknowledgments

We would like to thank Mr. Aoyama of Equipment Development Center of Institute for Molecular Science for his fabricating the three-dimensional magic mirror, and also thank staff members of UVSOR-II for their advice of the construction of the beamline. This work was partly supported by the Quantum Beam Technology Program of Japan Science and Technology Agency (2008–2013).

References

[1] Ex. WIRMS 2009 proceedings 2010 *AIP Conf. Proc.* **1214**.
[2] Nakazato T, Oyamada M, Niimura N, Urasawa S, Konno O, Kagaya A, Kato R, Kamiyama T, Torizuka Y, Nanba T, Kondo Y, Shibata Y, Ishi K, Ohsaka T and Ikezawa M 1989 *Phys. Rev. Lett.* **63** 1245.
[3] Carr G L, Kramer S L, Murphy J B, Lobo R P S M and Tanner D B 2001 *Nucl. Instrum. & Meth. Phys. Res. A* **463** 387.
[4] Byrd J M, Leemans W P, Loftsdottir A, Marcelis B, Martin M C, McKinney W R, Sannibale F, Scarvie T and Steier C 2002 *Phys. Rev. Lett.* **89** 224801.
[5] Feikes J, Holldack K, Kuske P, Wüstefeld G and Hübers H W 2004 *ICFA Beam Dynamics Newsletter* **35** p. 82.
[6] Byrd J M, Hao Z, Martin M C, Robin D S, Sannibale, Schoenlein R W, Zholents A A and Zolotorev M S 2006 *Phys. Rev. Lett.* **96** 164801.
[7] Singley E J, Abo-Bakr M, Basov D N, Feikes J, Guptasarma P, Holldack K, Hübers H W, Kuske P, Martin M C, Peatman W B, Schade U and Wšterfeld G 2004 *Phys. Rev. B* **69** 092512.
[8] Ortolani M, Lupi S, Baldassarre L, Schade U, Calvani P, Takano Y, Nagao M, Takenouchi T and Kawarada H 2006 *Phys. Rev. Lett.* **97** 097002.
[9] Takashima Y, Katoh M, Hosaka M, Mochihashi A, Kimura S and Takahashi T 2005 *Jpn. J. Appl. Phys.* **44** L1131.
[10] Shimada M, Katoh M, Kimura S, Mochihashi A, Hosaka M, Takashima Y, Hara T and Takahashi T 2007 *Jpn. J. Appl. Phys.* **46** 7939.
[11] Shimada M, Katoh M, Adachi M, Kimura S, Tanikawa T, Hosaka M, Yamamoto N, Takashima Y and Takahashi T 2009 *Phys. Rev. Lett.* **103** 144802.
[12] Bielawski S, Evain C, Hara T, Hosaka M, Katoh M, Kimura S, Mochihashi A, Shimada M, Szwaj C, Takahashi T and Takashima Y 2008 *Nat. Phys.* **4** 390.
[13] Evain C, Szwaj C, Bielawski S, Hosaka M, Takashima Y, Shimada M, Kimura S, Katoh M, Mochihashi A, Takahashi T and Hara T 2010 *Phys. Rev. ST Accel. Beams* **13** 090703.
[14] Labat M, Hosaka M, Mochihashi A, Shimada M, Katoh M, Lambert G, Hara T, Takashima Y and Couprie M E 2007 *Euro. Phys. J. D* **44** 187.
[15] Labat M, Hosaka M, Shimada M, Katoh M and Couprie M E 2010 *Phys. Rev. Lett.* **101** 164803.
[16] Tanikawa T, Adachi M, Zen H, Hosaka M, Yamamoto N, Taira Y and Katoh M 2010 *Appl. Phys. Express* **3** 122702.
[17] Kimura S, Nakamura E, Hosaka M, Takahashi T and Katoh M 2010 *AIP Conf. Proc.* **1234** 63.
[18] Kimura S, Nakamura E, Nishi T, Sakurai Y, Hayashi K, Yamazaki J and Katoh M 2006 *Infrared Phys. Tech.* **49** 147.
[19] Nasse M J, Reininger R, Kubala T, Janowski S and Hirschmugl C 2007 *Nucl. Instrum. & Meth. Phys. Res. A* **582** 107.
[20] Kimura S, Kimura H, Takahashi T, Fukui K, Kondo Y, Yoshimatsu Y, Moriwaki T, Nanba T and Ishikawa T 2001 *Nucl. Instrum. Meth. & Meth. Phys. Res. A* **467-468** 437.

Journal of Physics: Conference Series **359** (2012) 012010 doi:10.1088/1742-6596/359/1/012010

Photons and magnetic fields: The use of synchrotron sources to study pairbreaking in superconductors

Xiaoxiang Xi,[1] C Martin,[1] J Hwang,[1] RSPM Lobo,[1] H Tashiro,[1] H Zhang,[1] DH Reitze,[1] CJ Stanton,[1] DB Tanner[1] and GL Carr[2]

[1]Department of Physics, University of Florida,Department of Physics, University of Florida, Gainesville, Florida 32611, USA

[2]Photon Sciences, Brookhaven National Laboratory, Upton, New York 11973, USA

E-mail: `tanner@phys.ufl.edu`

Abstract. Pair-breaking effects in metallic superconductors have been studied by both linear and nonlinear spectroscopy using infrared synchrotron radiation. The measurements were performed at the National Synchrotron Light Source, Brookhaven National Laboratory, in magnetic fields up to 10 T. The optical conductivity of thin-film superconductors in applied magnetic fields has been estimated from the results of far-infrared transmission and reflection measurements. The combined measurements have been analyzed to give the real and imaginary parts of the conductivity. In turn, these quantities allow the magnetic-field dependence of the superconducting energy gap, pairbreaking parameter, and superfluid density to be estimated. Photons also may break Cooper pairs. Pump-probe studies of excess quasiparticle relaxation in the these superconducting films show a relaxation rate proportional to the excess quasiparticle number density, as expected for bimolecular recombination driven by a large excess quasiparticle population. Application of a magnetic field parallel to the sample surface is found to slow significantly the quasiparticle recombination process.

1. Introduction

Optical spectroscopy is a widely used technique for the study of the electronic properties of solids, with the interesting photon energy range extending at least from the far-infrared/terahertz range (sub-meV) to the ultraviolet (several eV). These experiments allow studies of physics as diverse as the energy gaps of metallic superconductors, spin excitations in magnets, free-carrier dynamics in metals and semiconductors, interband excitations, the vibrational modes of crystals, and spectral weight transfers associated with metal-insulator or superconducting phase transitions.

The great majority of optical studies are linear, equilibrium experiments. The only deviations from the ground state are caused by thermal excitations. In contrast, this paper describes studies in a rather different situation: infrared studies of dynamics of systems away from equilibrium, in some cases far from equilibrium. We investigate nonequilibrium states of matter through pump-probe far-infrared spectroscopy. A wide range of materials can be studied using this time-resolved method, which makes use of a key feature of synchrotron infrared radiation: the electrons in the synchrotron ring exist in tight bunches traveling around the ring, making the synchrotron light occur as pulses, emitted each time a bunch passes the aperture of the beamline. The length of the bunch sets the duration of the pulse, typically in the sub-nanosecond range. Although perhaps the least exploited aspect of synchrotron radiation, this time structure makes the synchrotron able to do a class of very important experiments that cannot be done in any other way: broadband infrared time-resolved or pump-probe spectroscopy.[1,2]

Our experimental setup uses a short-pulsed near-infrared laser as pump and the very broadband synchrotron infrared radiation as probe. Synchronization between laser pulses and infrared pulses gives the ability to study transient behavior on timescales from ~300 ps to ~100 ns. Spectrometers attached to the beamline permit spectroscopy of these transient events in the far-infrared to near infrared ranges.

2. Linear spectroscopy

2.1. Synchrotron infrared radiation

Our measurements used the far-infrared photons from beamline U4IR in the VUV synchrotron storage ring of the National Synchrotron Light Source. The spectra were collected using a Bruker IFS 66-v/S spectrometer and a high sensitivity, large area, B-doped Si composite bolometer operating at 1.8 K; cooled filters limited the upper frequency to 110 cm^{-1}. The samples were mounted in a ^{4}He Oxford cryostat equipped with a 10 Tesla superconducting magnet; the minimum temperature is 1.6 K.

The high brightness of the synchrotron radiation allowed transmission and reflection measurements to be made in the cramped space of the high field region after carrying it many meters from the synchrotron to the interferometer, on to the magnet, and eventually to the detector.

2.2. Magnetic-field-induced pair breaking in superconducting $Nb_{0.5}Ti_{0.5}N$

We measured the complex optical conductivity of a superconducting thin-film of $Nb_{0.5}Ti_{0.5}N$ in an external magnetic field. The field was applied parallel to the film surface and the conductivity extracted from far-infrared transmission and reflection measurements. The real part shows the superconducting gap, which we observe to be suppressed by the applied magnetic field. We compare our results with the pair-breaking theory of Abrikosov and Gor'kov[3] and confirm directly the theory's validity for the optical conductivity.

The normalized optical conductivities at 0, 5, and 10 T are shown in panels (a)–(c) of Fig. 1. A weak interference fringe in both the transmission and reflection measurements results in the excess σ_2/σ_N over the 40 to 80 cm^{-1} range. The solid lines are fits to the data using the pair-breaking theory as extended by Skalski *et al.*[4] to calculate σ_1/σ_N at 0 K. Panel (d) shows the fitted σ_1/σ_N at six different fields. The fits used the pair-breaking theory as extended by Skalski *et al.*[4] to calculate σ_1/σ_N at 0 K. Clearly, the absorption

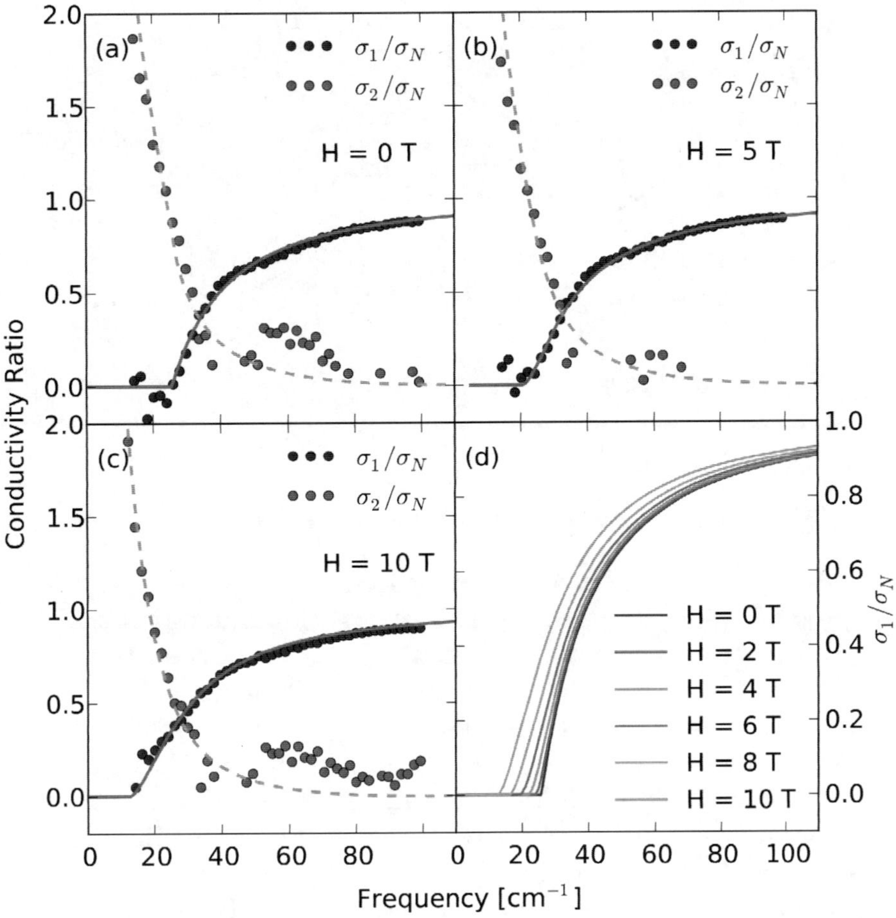

Fig. 1.(a)–(c): The real and imaginary parts of the $T = 3$ K superconducting state optical conductivity (normalized to the normal state conductivity) at three different applied magnetic fields. The solid lines are fits to σ_1/σ_N using the pair-breaking theory and the dashed lines show σ_2/σ_N from a Kramers-Kronig transform of the real part. (d): The fitted σ_1/σ_N at six fields.

edge moves to lower energy as the field increases. The field-induced pair breaking also smears out the gap-edge singularity in the quasiparticle density of states[4], so that the initial rise of σ_1 becomes less abrupt for increasing fields, as can be seen by comparing the 0 T and the 10 T results in Fig. 1. Our analysis[5] finds that the suppression of both the superconducting order parameter and the spectroscopic gap is in good accord with the pair-breaking theory.[3]

3. Pump-probe spectroscopy

3.1. Experimental methods

Figure 2 shows a schematic diagram of the pump-probe facility.[2] The sample is first irradiated with a short, relatively high-power, "pump" pulse, creating a nonequilibrium photoexcited state. These excitations, their interactions, and their dynamics modify the frequency-dependent optical properties of the material; these changes are detected using a "probe" pulse, which irradiates the material with a known time delay after the pump

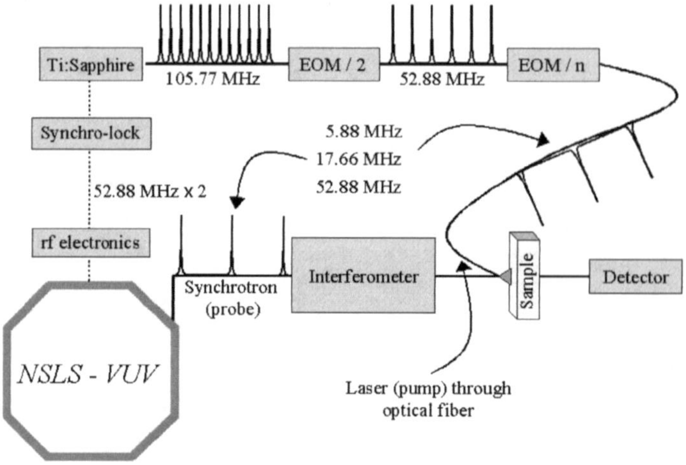

Fig. 2. Apparatus for time-resolved pump-probe measurements.

pulse. The evolution of these photoinduced changes in optical properties can be followed by varying the time delay. Beamline U4IR was used for pump-probe spectroscopy.

A major issue in any pump-probe spectroscopy is minimization of thermal effects due to the pump laser. If the experiment is done by measuring with the laser on for a while and then measuring with the laser off for a while, it is extremely difficult to ensure that the sample temperatures are identical for the two phases of the measurement. Because photoinduced effects are rather small, small temperature changes, hotter when illuminated and cooler when in the dark can mimic the nonequilibrium effects that one wishes to observe. Consequently in our experiment, the laser is *always* on. We inject a small audio-frequency modulation into the laser/synchrotron locking apparatus, causing the arrival time of the laser pulses to dither back and forth relative to the infrared pulse

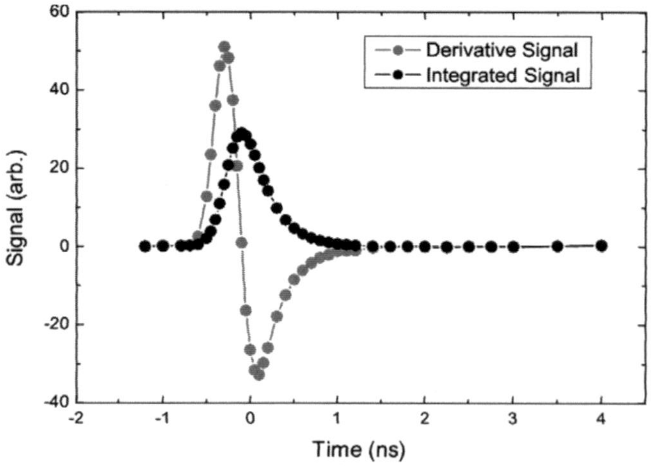

Fig. 3. The red curve shows the measured derivative signal and the black curve the integrated photoinduced absorption. The probe beam in this case is broad-band infrared synchrotron radiation, with energy over 10–80 cm^{-1}, peaked near the energy gap of the NbTiN film.

arrival. Just as when field modulations are used in an NMR or EPR experiment, this dither gives us a signal which is the derivative of the photoinduced response of the sample. Integration gives us the desired photoinduced signal, uncontaminated with thermal effects. Figure 3 shows the measured signal and its integral.

The photoinduced signal (at low fluence) is a convolution of the exponential relaxation of the photoinduced carriers and the near-Gaussian time dependence of the probe beam. We can fit to this behavior to extract the recombination time in the sample.

3.2. Quasiparticle dynamics

When a superconductor is subjected to a pulse of above-bandgap electromagnetic radiation, excess quasiparticles are created. Their number and how they recombine back into Cooper pairs are of fundamental importance. Calculations based on a phonon emission process,[6–8] found a rather fast recombination rate. However, Rothwarf and Taylor[7] showed that the measured (or effective) lifetime was longer than the "bare" or intrinsic recombination time due to a phonon-bottleneck effect.

Figure 4 shows a cartoon picture of the process. The laser photon creates a pair of very hot quasiparticles. These relax rapidly by interaction with other charged particles and by emission of high-energy phonons, breaking other Cooper pairs during this time. In picoseconds, much faster than the measurement capability of the synchrotron probe,[9] the quasiparticles relax to the gap edge. Here, each quasiparticle must find a partner of opposite spin to recombine into a Cooper pair. The recombination releases the pair binding energy in the form of a 2Δ phonon. However, this phonon can (and will) break another Cooper pair, creating two gap-edge quasiparticles. A stalemate occurs. Only if the phonon escapes from the superconductor or decays anharmonically can the excess quasiparticle density be relaxed. It is this bottleneck that we have studied.

Experiments on photoinduced pairbreaking include the work of Testardi,[10] who

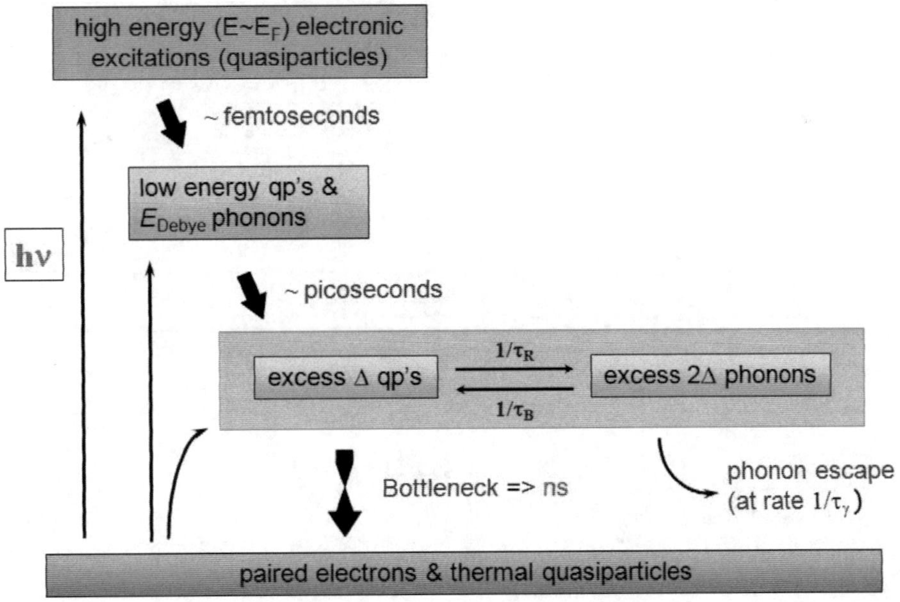

Fig. 4. Cartoon of the pairbreaking and recombination process, showing the bottleneck for the gap-edge quasiparticles and 2Δ phonons.

Journal of Physics: Conference Series **359** (2012) 012010 doi:10.1088/1742-6596/359/1/012010

showed that light of enough intensity could destroy the superconductivity in a process not attributable to lattice heating. Parker and Williams[11] and Schuller and Gray[12] used this phenomenon to measure pair recombination rates in illuminated tunnel junctions. Sai-Halasz *et al.*[13] used a pulsed laser to create the excess number of quasiparticles and analyzed the changes in the microwave reflectivity. Their data support the theory of Owen and Scalapino[14] for the low-fluence regime, where the number of quasiparticles created by the photon absorption is much smaller than the number of unpaired quasiparticles produced by finite sample temperature Johnson[15] was the first to use fast lasers and electronics to look at voltage transients with time resolution in the range of 100 ps. Measurements of quasiparticle recombination using synchrotron radiation include the work of Carr *et al.*[16] on Pb and Lobo *et al.*[17] on a variety of superconductors.

3.3. Pump-probe results

We used pump-probe spectroscopy to measure quasiparticle recombination dynamics in metallic superconductors ($Nb_{0.5}Ti_{0.5}N$ and NbN thin films) in an external magnetic field. The field was applied parallel to the film surface in order that vortex dynamics not play a role in the measurements. Figure 5 shows an example of our results. The photoinduced signal—proportional to the excess quasiparticle density; $S(t) \propto N_{ex}(t)$—is shown versus the delay time between laser pump and far-infrared probe. Note the semilog presentation. The left panel shows the relaxation at low laser fluence whereas the right panel shows the behavior at high fluence. The low fluence data are close to a simple exponential; the high fluence data are not. At high fluences, the recombination is fast, and then slows later in the process, as the excess density becomes smaller. Note that the signal is followed over two orders of magnitude in its strength.

The effect of field is to decrease the rate in both low and high fluence situations. The recovery takes longer when the field strength is increased. We can consider two

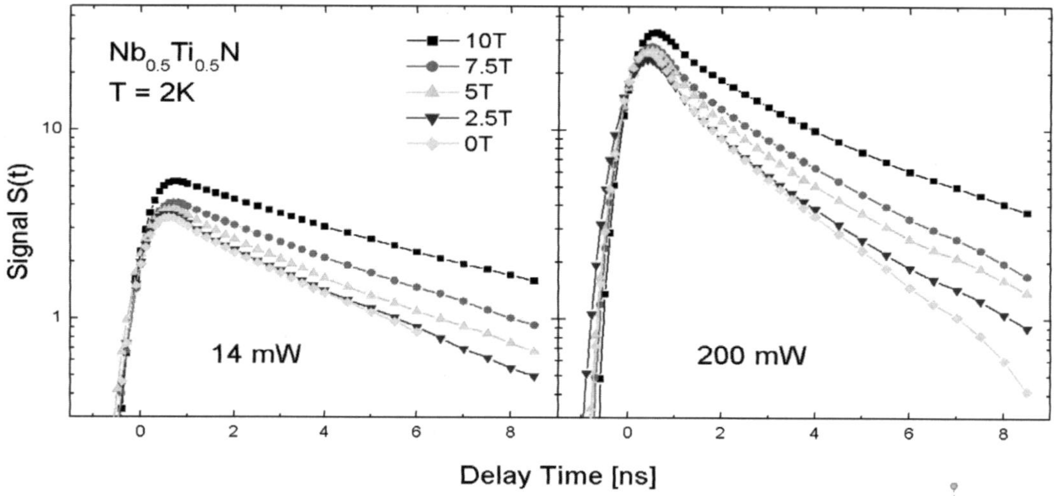

Fig. 5. The photoinduced signal is shown on a logarithmic scale versus the delay time between laser pump and far-infrared probe. The left panel shows the signal for low laser fluence and the right panel for higher fluence. The measurements were made with the sample immersed in 2 K liquid Helium and at applied magnetic fields from 0 to 10 T.

effects to understand this result. In one, the magnetic field produces a paramagnetic spin polarization, making it difficult for a majority-spin quasiparticle to find an opposite-spin partner to pair up with, thus slowing the recombination. Although attractive, the presence of spin-orbit coupling reduces the importance of this effect. The second effect is finite pair lifetime, observed in the linear spectroscopy in field (Fig. 2). We can therefore understand the recombination within the same pairbreaking theory[3] used to explain the linear spectroscopy.

Acknowledgements

This research has been supported by the DOE through contracts DE-FG02-02ER45984 at Florida and DE-AC02-98CH10886 at BNL.

References

1. Ederer D L, Rubensson J E, Mueller D R, Shuker R, O'Brien W L, Jai J, Dong Q Y, Calcott T A, Carr G L, Williams G P, Hirschmugl C J, Etemad S, Inam A, and Tanner D B 1992 *Nucl. Instrum. Methods* **A319** 250
2. Lobo R P S M, Carr G L, LaVeigne J, Reitze D H, and Tanner D B 2002 *Rev. Sci. Instrum.* **73** 1
3. Abrikosov A A and Gor'kov L P 1961 *Soviet Phys. JETP* **12** 1243
4. Skalski S, Betbeder-Matibet O, and Weiss P R 1964 *Phys. Rev.* **136** A1500
5. Xi Xiaoxiang, Hwang J, Martin C, Tanner D B, and Carr G L 2010 *Phys. Rev. Lett.* **105** 257006
6. Schrieffer J R and Ginsberg D M 1962 *Phys. Rev. Lett.* **8** 207
7. Rothwarf Allen and Cohen Michael 1963 *Phys. Rev.* **130** 1401; Rothwarf Allen and Taylor B N 1967 *Phys. Rev. Lett.* **19** 27
8. Gray K E 1969 *Philos. Mag.* **20** 267
9. The minimum pulse duration in the NSLS VUV ring is ~350 ps. In the NSLS II, presently under construction, this timescale is expected to reach 10 ps.
10. Testardi L R 1971 *Phys. Rev. B* **4** 2189
11. Parker W H and Williams W D 1972 *Phys. Rev. Lett.* **29** 924
12. Schuller Ivan and Gray K E 1976 *Phys. Rev. Lett.* **36** 429
13. Sai-Halasz G A, Chi C C, Denenstein A, and Langenberg D N 1974 *Phys. Rev. Lett.* **33** 215
14. Owen C S and Scalapino D J 1972 *Phys. Rev. Lett.* **28** 1559
15. Johnson Mark 1991 *Phys. Rev. Lett.* **67** 374
16. Carr G L, Lobo R P S M, LaVeigne J, Reitze D H, and Tanner D B 2000 *Phys. Rev. Lett.* **85** 3001
17. Lobo R P S M, LaVeigne J D, Reitze D H, Tanner D B, Barber Z H, Jacques E, Bosland P, Burns M J, and Carr G L 2005 *Phys. Rev. B* **72** 024510

Journal of Physics: Conference Series **359** (2012) 012011 doi:10.1088/1742-6596/359/1/012011

Enroute to investigating protein dynamics under selective vibrational excitation at the THz FEL FELBE

C Bauer[1,3*], M Gensch[1,2] and J Heberle[3]
[1] Institute of Radiation Physics, Helmholtz-Zentrum Dresden-Rossendorf, Dresden, Germany
[2] Institute of Ion Beam Physics and Materials Research, Helmholtz-Zentrum Dresden-Rossendorf, Dresden, Germany
[3] Experimental Molecular Biophysics, Department of Physics, Freie Universität Berlin, Germany

E-mail: c.bauer@hzdr.de

Abstract. We aim at investigating proteins under irradiation with intense THz radiation tuned into resonance to specific vibrational modes. This approach is much in analogy to recent experiments that showed selective vibrational control in Complex materials [1, 2, 3]. To achieve the necessary sensitivity for protein dynamics we combine a novel time-resolved IR difference spectroscopic setup with uniquely intense, tuneable narrow bandwidth THz radiation (1.2 – 75 THz) of the free electron laser FELBE.

1. Introduction

At the THz free electron laser FELBE [4] a novel THz pump FT-IR step scan probe instrumentation shall be installed. This unique set up pairs the peak power THz pulses from the FEL with the enormous sensitivity of time resolved FTIR difference spectroscopy. This set up shall be employed for investigations into selective vibrational control of protein reactions. We believe the capability of FT-IR difference spectroscopy to detect changes in the environment of one single bond [5] will prove crucial for this field.

Pilot experiments shall be performed on photosensitive proteins, namely bacteriorhodopsin. Organisms are using photosensitive proteins to optimize their living conditions to the environment. One of the model photoreceptor proteins in the focus of investigation over the last decades is bacteriorhodopsin (bR). Bacteriorhodopsin is the sole protein of the purple membrane of the archaebacterium *Halobacterium salinarum* and belongs to the family of seven-helical trans-membrane proteins. Upon illumination, its chromophore retinal isomerizes around the C_{13}-C_{14} double bond and bR pumps a proton from the cytoplasmic to the extracellular side. This proton gradient is used by the bacterium to drive photosynthetic ATP production under low oxygen level. After excitation of the light-adapted state, the protein undergoes a sequence of intermediate states returning to the initial state, which is called photocycle. These short-lived intermediate states (femtosecond to millisecond) show different absorption properties in the UV/Vis and, most importantly, in the infrared (IR) spectral range [6, 7].

It was shown earlier that the initial step of the photocycle of retinal can be influenced under coherent control [8]. Prokhorenko *et al.* could demonstrate an optimized isomerization yield of retinal using femtosecond laser pulse trains in the near infrared. These pulse trains have a temporal modulation

(caused by an acousto-optic modulator using an evolutionary algorithm) with a period of 145 femtoseconds, which corresponds to a frequency of 6.89 THz.
In our pilot experiment, we want to investigate whether control of the isomerization of retinal by direct selective excitation of the related torsional modes in the vicinity of the C_{13}-C_{14} double bond of retinal can be achieved. In case of success we envision that the use of direct excitation by narrow bandwidth THz radiation may allow an increased level of control. Employing pump radiation in the THz regime, we avoid multi-photon excitation, which occur using high peak-power femtosecond laser pulses and likely limit the degree of control to be achieved in coherent control experiments [9].

1. Vis-THz Pump IR Difference Spectroscopy Probe Set-Up At FELBE

To probe the effect of the narrow bandwidth THz radiation on protein reactions, the well-established method of IR difference spectroscopy [10] shall be used. This method is a very specific technique that can sensitively probe intra- and inter-molecular effects by monitoring the correlated modulation of the corresponding infrared spectra. During a reaction only certain parts of the proteins undergo molecular changes and their identification in the absolute IR spectra is impeded by the strong vibrational background of the whole protein. This obstacle can be overcome by recording the IR difference between two reaction states which renders IR spectroscopy selective. The difference spectra exhibits bands characteristic only for the molecular groups that are changing during the protein activity. Since the resulting absorbance changes are usually very small, of the order of 10^{-4} or less, their detection requires accurate and very sensitive instrumentation. The setup [11] is currently installed in one dedicated laboratory at FELBE where a FT-IR spectrometer (Vertex 80v from Bruker) with step-scan option is combined with one of the FEL beamlines (see figure 1). The issue of different penetration depth of the Vis, THz and MIR radiation involved in the pilot experiment is overcome by a novel ATR set-up [12] (see figure 2), where the two pump beams, THz tunable between 1.2 and 75 THz from the FEL and 532 nm radiation from a table-top frequency-doubled Nd:YAG laser system irradiate the bR film from the top while the broadband infrared radiation from the globar source of the FT-IR spectrometer is probing the IR absorption by attenuated total reflection from inside the ATR prism.

2. Summary and Outlook

A novel three-color pump probe set up is currently under development at the THz free electron laser FELBE in Dresden. To our knowledge this will be the first instrumentation that allows state of the art step-scan FT-IR difference spectroscopy to be performed on proteins under illumination of highly intense narrow band THz FEL radiation. Our efforts are driven by the quest to verify if protein reactions can be controlled by selective vibrational excitation of relevant low energy modes. As a model system, we have chosen bR which has been shown earlier that isomerization dynamics can be manipulated by the related approach of so called coherent control [8].

Acknowledgements

We would like to thank the Bundesministerium für Bildung und Forschung for funding through the PIDID proposal # 05K10KEB. C.B. would like to thank S. Winnerl and W. Seidel for technical assistance.

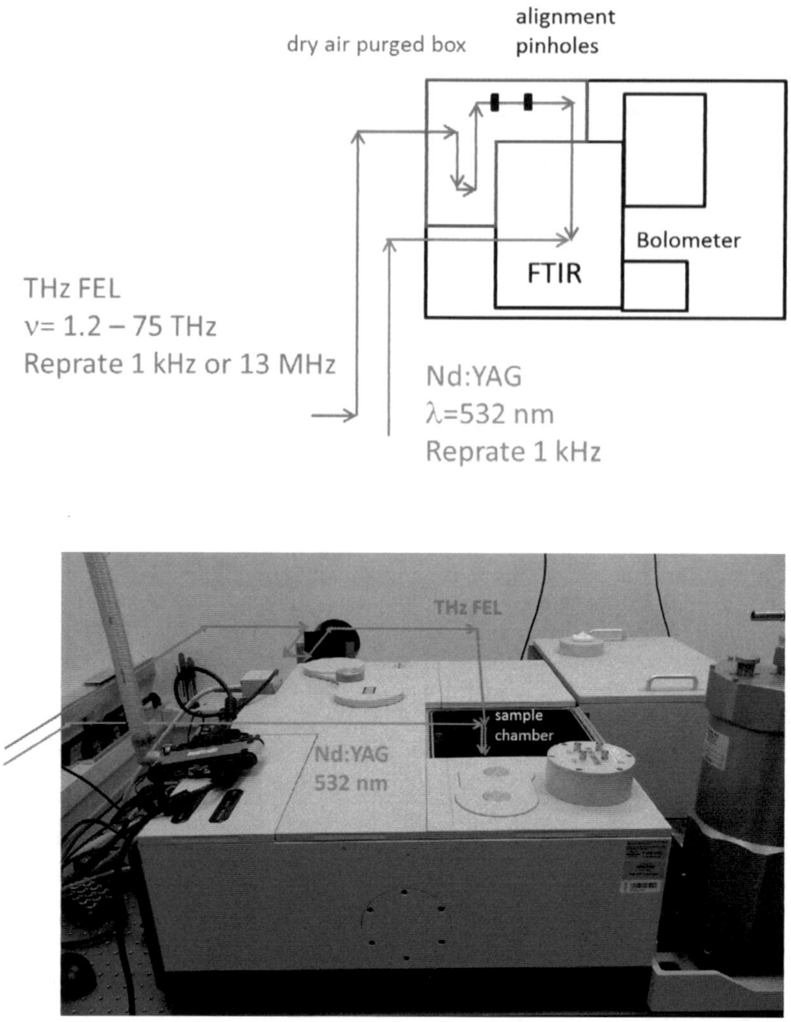

Figure 1:
Scheme of the experimental setup using a Bruker 80v FT-IR spectrometer combined with the THz FEL as the second pump source (top). The photo (bottom) shows our spectrometer and the pathways of the THz FEL and the visible Nd:YAG pump laser at 532 nm.

Journal of Physics: Conference Series **359** (2012) 012011 doi:10.1088/1742-6596/359/1/012011

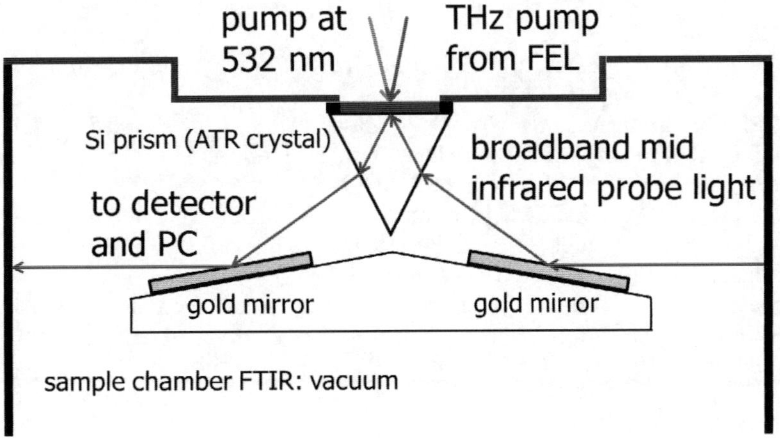

Figure 2:
Scheme of the single reflection ATR (attenuated total reflection) setup that will be applied for pump-probe experiments [12]. The two pump beams irradiate the sample from the top. The protein is probed by making use of an attenuated total reflection prism where the protein is deposited on to the surface of a highly refractive crystal material. The infrared probe beam is focused into the crystal at an angle above the critical angle for total internal reflection. At the interface with the rarer medium a non-radiative evanescent field evolves which decays exponentially outside the crystal and allows thereby measurements with a smaller effective penetration depth.

References

[1] Rini M, Tobey R, Dean N, Itatani J, Tomioka Y, Tokura Y, Schoenlein R W, Cavalleri A 2007 *Nature* **449** 72-74
[2] Tobey R I, Prabhakaran D, Boothroyd A T, Cavalleri A 2008 *Phys. Rev. Lett.* **101** 197404
[3] Fausti D, Tobey R I, Dean N, Kaiser S, Dienst A, Hoffmann M C, Pyon S, Takayama T, Takagi H, Cavalleri A 2011 *Science* **331** 189-191
[4] Zvyagin S A, Ozerov M, Cizmar E, Kamenskyi D, Zherlitsyn S, Herrmanssdörfer T, Wosnitza J, Wünsch R, Seidel W 2009 *Rev. Sci. Instrum.* **80** 073102
[5] Heberle J, Gensch T, 2001 *Nature Struct. Biol.* **8** 195-197
[6] Sass H J, Schachowa I W, Rapp G, Koch M H, Oesterhelt D, Dencher N A, Büldt G 1997 *EMBO Journal* **16** 1484-1491
[7] Zscherp C, Heberle J 1997 *J. Phys. Chem B* **101** 10542-10547
[8] Prokhorenko V I, Nagy A M, Waschuk S A, Brown L S, Birge R R, Miller R J D 2006 *Science* **313** 1257-1261
[9] Prokhorenko V I, Halpin A, Johnson P J M, Miller R J D, Brown L S 2011 *J. Chem. Phys.* **134** 085105
[10] Mäntele W, 1993 *Trends Biochem. Sci.* **18** 197-202
[11] Radu I, Schleeger M, Nack M, Heberle J 2011 *Aust. J. Chem* **64** 9-15
[12] Bauer C, to be published

Journal of Physics: Conference Series **359** (2012) 012012 doi:10.1088/1742-6596/359/1/012012

Volume dependent quasiparticle spectral weight in $NiS_{2-x}Se_x$ system.

C. Marini[1,2], A. Perucchi[2,3], P. Dore[4], D. Topwal[5], D.D. Sarma[5], S. Lupi[2], and P. Postorino[2]

[1] European Synchrotron Radiation Facility, 6 Rue Jules Horowitz, BP220, 38043 Grenoble Cedex, France
[2] CNR-IOM and and Dipartimento di Fisica, Università di Roma "La Sapienza", Piazzale Aldo Moro 2, I-00185 Roma, Italy
[3] Sincrotrone Trieste S.C.p.A., Area Science Park, I-34012 Basovizza, Trieste, Italy
[4] CNR-SPIN and and Dipartimento di Fisica, Università di Roma "La Sapienza", Piazzale Aldo Moro 2, I-00185 Roma, Italy
[5] Solid State and Structural Chemistry Unit,Indian Institute of Science, Bangalore 560012, India

E-mail: carlo.marini@esrf.fr

Abstract. We discuss the evolution of Infrared reflectivity at room temperature under various pressures (P) and Se alloying concentration in the strongly correlated $NiS_{2-x}Se_x$ pyrite. Measurements gave a complete picture of the optical response of the system on approaching the P-induced and Se-induced metallic state. A peculiar non-monotonic (V-shaped) volume dependence was found for the quasiparticle spectral weight of both pure and Se-doped compounds.

1. Introduction

In the past sixty years, much progress has been made from both theoretical and experimental sides in understanding strongly correlated electrons and metal insulator transition (MIT)[1]. Cubic $NiS_{2-x}Se_x$ series provided an important model system for a systematic investigation, exhibiting a MIT controlled by varying the Se content x, or pressure P[2, 3, 4].

The two end members of the family, NiS_2 and $NiSe_2$, are charge transfer (CT) insulators [5] and metals[6] respectively; they can be mixed in all proportions, thereby permitting a graded variation in electrical and magnetic properties through adjustment of the stoichiometry ratio x: $NiS_{2-x}Se_x$ exhibits first an anti-ferromagnetic insulator ground state for zero and low Se-content and then it enters an anti-ferromagnetic metallic phase around x = 0.40. A paramagnetic metallic ground state is finally achieved at room temperature around x = 0.60 [7].

An alternative way to induce a metallic state in NiS_2 is applying a hydrostatic P. A P-induced MIT was observed in pure NiS_2 for P > 4GPa[4, 9, 10, 11]. Whereas chemical substitution inevitably introduces impurities, changing many physical parameters in an incontrollable way, P offers a more systematic way of tuning the properties.

Infrared spectroscopy is a powerful tool to study the properties of a system since the optical response of the free-charge carriers is determined by the frequency-dependent complex conductivity. Recently, we carried out an infrared (IR) investigation on several compounds of the $NiS_{2-x}Se_x$ pyrite family (i.e. x = 0.00, 0.55, 0.60, and 1.20)[10]. In this paper, we present the new high P (HP) IR data on the x = 0.55 sample re-discussing the already published ones, thus obtaining a deeper insight into low energy electron dynamics in the system.

2. Experimental methods

The measurements were performed at the high brightness infrared synchrotron-radiation source SISSI@Elettra (Trieste)[12], using a cassegrain-based Hyperion 2000 IR microscope equipped with a MCT detector and coupled to a Bruker IFS 66v interferometer. A screw clamped opposing-plate diamond-anvil cell (DAC) equipped with 800 μm culet II A diamonds has been used for IR HP experiments. Pressure has been determined by ruby fluorescence technique[13]. Thick Sample slabs have been obtained by pressing finely milled sample powder between the diamond anvils. The sample slabs have been placed on top of a pre-sintered KBr pellet in the gasket hole[14]. This procedure ensures a clean sample-diamond interface, where reflectivity spectra, $R_{sd}(\omega)$, have been measured. Further details on the measurement procedures are reported elsewhere[10, 15].

3. Results and discussion

Room temperature $NiS_{2-x}Se_x$ reflectivity curves $R(\omega)$ were measured as a function of x (i.e. x = 0.00, 0.55, 0.60, and 1.20) between 50 and 35 000 cm^{-1} on well characterized high-density pellets. The $R(\omega)$ of NiS_2 at ambient P, shown in panel (a) of Fig. 1, is nearly flat from 50 to 10 000 cm^{-1} except for weak phonon contributions at 260 and 290 cm^{-1}. On increasing the Se content, $R(\omega)$ is progressively enhanced at low frequencies, characteristic of a metallic behavior. $R_{sd}(\omega)$ as function of P for the pure and x = 0.55 compounds are shown in panel (b) and (c) respectively. Zero pressure reflectance spectrum, as previously evaluated at the diamond-sample interface[10], is also shown for comparison. In the parent compound on increasing the pressure, $R_{sd}(\omega)$ progressively increases at low frequency showing an overdamped behavior, as a signature of a correlated bad metallic state. At high frequencies all $R_{sd}(\omega)$ converge above 10 000 cm^{-1}.

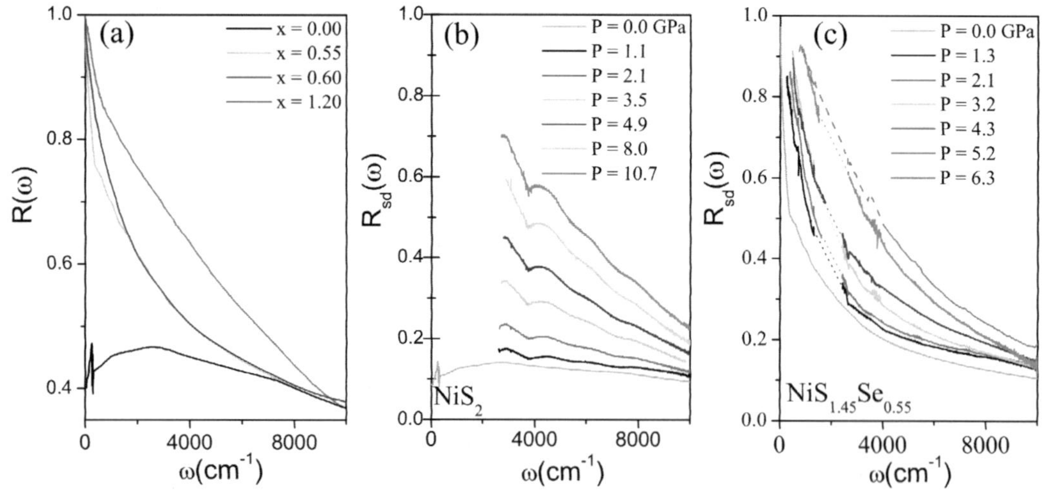

Figure 1. (a) ambient conditions $R(\omega)$ for several compounds of the $NiS_{2-x}Se_x$ pyrite family (i.e. x = 0.00, 0.55, 0.60, and 1.20); NiS_2 (b) and $NiS_{1.45}Se_{0.55}$ (c) $R_{sd}(\omega)$ as function of P.

Journal of Physics: Conference Series **359** (2012) 012012 doi:10.1088/1742-6596/359/1/012012

On increasing P in Se-substituted sample a similar $R_{sd}(\omega)$ behavior is observed. Also in this case the progressive enhancement of the $R_{sd}(\omega)$ suggests the crossover between insulator and metallic states.

Reflectivity spectra have been analyzed through a Drude-Lorentz model with three oscillators: a Drude term, a broad mid-infrared contribution centered around 2000 cm^{-1}, and an intense high frequency contribution centered around 10000 cm^{-1} which mimic the quasi-particle (QP) coherent excitations around the Fermi Energy and QP transitions from the QP peak to the upper and lower Hubbard bands respectively [10]. X and P effects can be thus enlightened by an analysis of the spectral weight (SW), defined as the integrated area of Drude plus mid-IR components. On increasing the Se content a linear increase of SW is observed, as shown in panel (a) of fig. 2. We turn now to the high-P measurements. For the both parent and Se substituted compound SW is nearly zero at ambient-P and then increases reflecting the progressive enhancement of the electron kinetic energy corresponding to the metallization of the system. A change of the slope is thus observed above 5 and 2 GPa respectively.

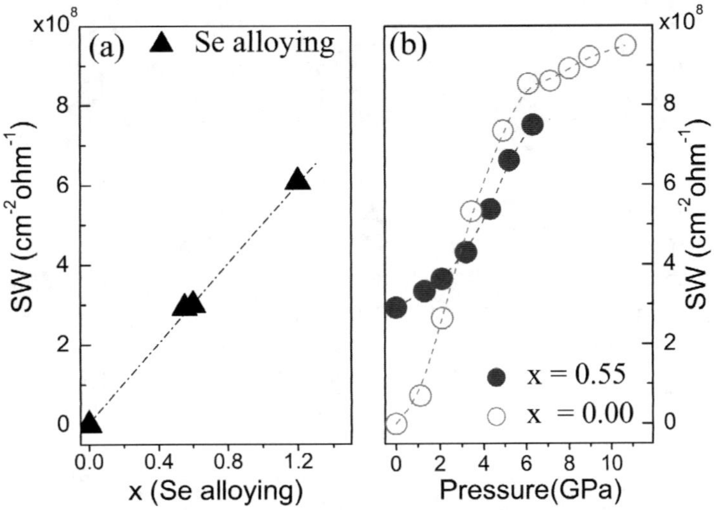

Figure 2. Quasi-particle spectral weight of $NiS_{2-x}Se_x$ as function of x (a) and as function of P (b) for x = 0.00 and x = 0.55 Se substituted $NiS_{2-x}Se_x$ pyrite.

4. Conclusions

The measurements presented here together with those reported in Ref. [10] provide data on the dependence of the quasi-particle spectral weight on pressure P and Se alloying of $NiS_{2-x}Se_x$ system. All the data can be plotted together by considering their dependence from the volume, bearing in mind that V is compressed by P while it is expanded by x. For the volume expansion given by Se alloying, X-ray diffraction data are available in the literature[16], while for the volume compression given by P structural data are only available for NiS_2 up to 6.5 GPa[17]. However,a reliable estimation of V at higher P values for NiS_2 and for all the pressures in Se-substituted compounds can be obtained by using the procedure previously developed in ref. [18]. Briefly, through the specific-heat data[19, 20] it is possible to estimate the bulk modulus of the system. The Murnagham Equation[21] finally provides the pressure dependence of V. This procedure gives volume values in very good agreement with those resulting from LDA calculations for NiS_2 up at least 10 GPa[10].

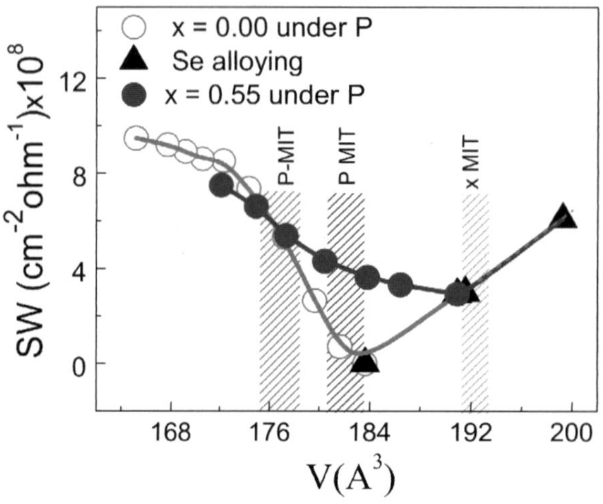

Figure 3. Quasi-particle spectral weight of $NiS_{2-x}Se_x$ as a function of volume: red open circles x=0.00, blue full circles x=0.55 (present data), triangles ambient condition x=0.00, 0.55, 0.60, and 1.20. The hatched stripe mark the volume region where the MITs occurs (see the text).

The volume dependence of SW is reported in Fig. 3. In pure NiS_2, SW slightly decreases on increasing V up to$\sim$ 173 A^3, i.e. on releasing P from highest value down to $\sim$ 5 GPa, and then drops to the minimum value on further increasing V up to 183 A^3, i.e. to the volume of NiS_2 at ambient P. Finally on entering the Se alloying regime (i.e. V $>$ 183 A^3) the SW restarts to raise, due to the decrease of the charge transfer gap. This originates a characteristic V shape behaviour. As to x=0.55 compound, SW decreases more slowly on increasing the volume, with a change in slope at $\sim$ 183 A^3 (i.e. at $\sim$ 2 GPa), and reaches the minimum value around 191 A^3, corresponding to the volume at ambient P of $NiS_{1.45}Se_{0.55}$. On further increasing V a new metallization occurs (due to the charge gap closure). The main effect of Se substitution results in a less pronounced V shape behavior. This is because the p orbitals of the Se^{2-} ion have larger spatial extent than those of S^{2-} and hence larger p-d and p-p transfer interactions, resulting in a less pronounced insulating character, as also suggested by the greater SW value at zero P.

In summary, a detailed IR study on the P- and x- induced MITs in $NiS_{2-x}Se_x$ pyrite has been presented. A strong non monotonic behaviour of the quasiparticle SW as function of the volume is observed and discussed. Present results open to new experimental and theoretical quests aimed at addressing the effects of increasing electron-electron interaction by appropriate changes in the unit cell volume in strongly correlated system.

References

[1] Imada H, Fujimori A and Tokura Y 1998 *Rev. Mod. Phys.* **70** 1039.
[2] Kwizera P, Dresselhaus M S and Adler D 1980 *Phys. Rev.* B **21** 2328.
[3] Matsuura M, Hiraka H, Yamada K and Endoh Y 2000 *J. Phys. Soc. Jpn.* **69** 1503.
[4] Miyasaka S, Takagi H, Sekine Y, Takahashi H, Mori N and Cava R J, 2000 *J. Phys. Soc. Jpn.* **69** 3166.
[5] Zaanen J, Sawatzky G A and Allen J W 1985 *Phys. Rev. Lett.* **55** 418.
[6] Yao X, Honig J M, Hogan T, Kannewurf C and Spalek J 1996 *Phys. Rev.* B **54** 17469.
[7] Iwaya K, Kohsaka Y, Satow S, Hanaguri T, Miyasaka S and Takagi H 2004 *Phys. Rev.* B **70**, 161103(R).
[8] Mott N F 1949 *Proc. Phys. Soc.* A **64** 416.
[9] Niklowitz P G, Steiner M J, Lonzarich G G, Braithwaite D, Knebel G, Flouquet J and Wilson J A 2008 *Phys. Rev.* B **77** 115135.
[10] Perucchi A, Marini C, Valentini M, Postorino P, Sopracase R, Dore P, Hansmann P, Jepsen O, Sangiovanni G, Toschi A, Held K, Topwal D, Sarma D D and Lupi S 2009 *Phys. Rev.* B **80** 073101.

[11] Kunes J, Baldassarre L, Schochner B, Rabia K, Kuntscher C A, Korotin Dm M, Anisimov V I, McLeod J A, Kurmaev E Z and Moewes A 2009 *Phys. Rev.* B **81** 035122.

[12] Lupi S, Nucara A, Perucchi A, Calvani P, Ortolani M, Quaroni L and Kiskinova M 2007 *J. Opt. Soc. Am.* B **24** 959.

[13] Mao H K, Bell P M, Shaner J W and Steinberg D J 1978 *J. Appl. Phys.* **49**, 3276.

[14] Sacchetti A, Cestelli Guidi M, Arcangeletti E, Nucara A, Calvani P, Piccinini M, Marcelli A and Postorino P 2006 *Phys. Rev. Lett.* **96** 035503.

[15] Marini C, Arcangeletti E, Di Castro D, Baldassare L, Perucchi A, Lupi S, Malavasi L, Boeri L, Pomjakushina E, Conder K and Postorino P 2008 *Phys. Rev.* B **77** 23511.

[16] De las Heras C and Agull-Rueda F 2000 *J. Phys. Condens. Matter* **12** 5317.

[17] Fujii T, Tanake K, Marumo F and Noda Y 1987 *Mineral. J.* **13** 448.

[18] Sacchetti A, Arcangeletti E, Perucchi A, Baldassarre L, Postorino P, Lupi S, Ru N, Fisher I R and Degiorgi L 2007 *Phys. Rev. Lett.* **98**, 026401.

[19] Sudo S, Nishioka T, Miyako Y, and Miyadai T 1986 *J. Phys. Soc. Jpn.* **55** 1806.

[20] Yao X, Kuo Y K, Powell D K, Brill J W, and Honig J M 1997 *Phys. Rev.* B **56** 7129.

[21] Murnaghan F D 1944 *Proc. Natl. Acad. Sci. U.S.A.* **30** 244.

Journal of Physics: Conference Series **359** (2012) 012013 doi:10.1088/1742-6596/359/1/012013

A simple method for the Kramers-Kronig analysis of reflectance spectra measured with diamond anvil cell

H. Okamura

Department of Physics, Graduate School of Science, Kobe University, Kobe 657-8501, Japan

E-mail: okamura@kobe-u.ac.jp

Abstract. When the optical reflectance spectrum of a sample under high pressure is studied with a diamond anvil cell (DAC), it is measured at a sample/diamond interface. Due to the large refractive index of diamond, the resulting reflectance $R_d(\omega)$ may substantially differ from that measured in vacuum. To obtain optical constants from $R_d(\omega)$, therefore, the usual Kramers-Kronig (KK) transform cannot be straightforwardly applied, and either a spectral fitting or a modified KK transform has been used. Here we describe an alternative method to perform KK analysis on $R_d(\omega)$. This method relies on the usual KK transform with an appropriate cutoff and extrapolation to $R_d(\omega)$, and may offer a simpler approach to obtain infrared conductivity from measured $R_d(\omega)$.

1. Introduction

Infrared (IR) spectroscopy has been a powerful tool to study the microscopic carrier dynamics and electronic structures in strongly correlated electron materials, such as rare earth (f electron), transition metal (d electron), and organic (p electron) compounds [1]. The IR spectroscopy technique has been also performed under high pressure using a diamond anvil cell (DAC) [2-13] since the strongly correlated materials show many interesting physical properties under high pressure. In a DAC, a pair of diamond anvils and a thin metal gasket are used to seal a sample and a pressure transmitting medium [6]. A typical diameter of the diamond surface is 0.8 mm to reach a pressure of 10 GPa, and 0.6 mm to reach 20 GPa. Therefore the sample in this experiment should have dimensions of the order of 100 μm. To perform an infrared (IR) reflectance study on such a small sample under the restricted sample space in a DAC, synchrotron radiation (SR) has been used as a bright source of both far and mid-infrared. In fact, high pressure IR spectroscopy with DAC is currently one of the major applications of IR-SR [4-7,9-13].

With a DAC, the reflectance is measured between the sample/diamond interface, in contrast to the usual case of sample/vacuum or sample/air interface. The normal-incidence reflectance of a sample relative to a transparent medium of (real) refractive index n_0 is given by the Fresnel's formula as [14, 15]:

$$R(\omega) = \frac{(n-n_0)^2+k^2}{(n+n_0)^2+k^2}. \qquad (1)$$

Here, $\hat{n} = n + ik$ is the complex refractive index of the sample, and n_0=2.4 for diamond and 1.0 for vacuum. Hereafter, we denote $R(\omega)$ at sample/diamond interface as $R_d(\omega)$, and that at sample/vacuum interface as $R_0(\omega)$. From Eq. (1), it is easily seen that $R_d(\omega)$ of a sample measured in DAC may be substantially different from $R_0(\omega)$.

Journal of Physics: Conference Series **359** (2012) 012013 doi:10.1088/1742-6596/359/1/012013

The purpose of this study is to consider the Kramers-Kronig (KK) analysis of $R_{\rm d}(\omega)$ data measured in DAC. KK analysis has been widely used to derive optical constants such as the refractive index, dielectric function and optical conductivity from a measured $R_0(\omega)$ spectrum [14, 15]. However, due to the difference between $R_0(\omega)$ and $R_{\rm d}(\omega)$ discussed above, the usual KK analysis method cannot be straightforwardly applied to $R_{\rm d}(\omega)$ [16]. To derive optical constants from $R_{\rm d}(\omega)$, therefore, previous high pressure IR studies used either a Drude-Lorentz spectral fitting [2, 5, 6, 7, 9, 11, 12] or a modified KK transform [3, 8, 10, 13]. In this work, we propose a different method, which relies on the usual KK transform with an appropriate cutoff to the $R_{\rm d}(\omega)$, as an alternative approach to obtain the infrared $\sigma(\omega)$ from $R_{\rm d}(\omega)$. The validity of the proposed method is demonstrated with actually measured reflectance data of $PrRu_4P_{12}$.

2. Kramers-Kronig analysis of reflectance spectra

The complex reflectivity of the electric field, $\hat{r}$, is expressed as [14, 15]

$$\hat{r}(\omega) = \frac{n_0 - \hat{n}(\omega)}{n_0 + \hat{n}(\omega)} = r(\omega)e^{i\theta(\omega)}. \tag{2}$$

Here $r(\omega)$ is the square root of the reflectance $R(\omega)$, which is actually measured in experiments. Then the real and imaginary parts of $\hat{n}$ can be expressed in terms of r and θ as

$$n = \frac{1 - r^2}{1 + r^2 + 2r\cos\theta} \cdot n_0 \tag{3}$$

and

$$k = \frac{-2r\sin\theta}{1 + r^2 + 2r\cos\theta} \cdot n_0, \tag{4}$$

respectively. Therefore, if $\theta(\omega)$ can be derived from measured $r(\omega)$ with KK analysis even for the sample/diamond reflection case, $n(\omega)$ and $k(\omega)$ can also be derived simply by setting n_0=2.4 in Eqs. (2) and (3). Then, the imaginary part of the complex dielectric function is given as $\epsilon_2 = 2n(\omega)k(\omega)$, and the optical conductivity is given as $\sigma(\omega) = \frac{\omega}{4\pi}\epsilon_2(\omega)$ [14, 15].

In performing KK analysis on reflectance data, usually the logarithm of $\hat{r}$, namely

$$\ln\hat{r}(\omega) = \ln r(\omega) + i\theta(\omega) \tag{5}$$

is regarded as a complex response function. In the case of sample/vacuum reflection, the KK relation between $\ln r(\omega)$ and $\theta(\omega)$ is expressed as [14, 15]

$$\theta(\omega) = -\frac{2\omega}{\pi} P\int_0^\infty \frac{\ln r(\omega')}{\omega'^2 - \omega^2} d\omega'. \tag{6}$$

Here, P denotes the principal value. In deriving this relation, it is required that $\ln\hat{r}(\hat{\omega})$ has no poles in the upper complex $\hat{\omega}$ plane when $|\omega|$ is finite. This is correct since $r(\omega) \to 0$ only when $\omega \to \infty$ in the case of n_0=1. However, when $n_0 > 1$ as in the case of sample/diamond interface, $\hat{n} = n_0$ may be satisfied at some point on the upper imaginary axis [16]. This point is denoted as $\hat{\omega} = i\beta$, where β is a real, positive and finite number. When $\hat{n} = n_0$, $\hat{r} = 0$ from Eq. (2) and $\ln\hat{r}$ therefore has a pole at $\hat{\omega} = i\beta$. Accordingly, the KK relation in this case must be modified to [16]

$$\theta(\omega) = -\frac{2\omega}{\pi} P\int_0^\infty \frac{\ln r(\omega')}{\omega'^2 - \omega^2} d\omega' + [\pi - 2\arctan(\beta/\omega)]. \tag{7}$$

Namely, the presence of a medium with $n_0 > 1$ brings an extra phase shift, indicated by the square bracket in Eq. (7), into the KK relation. Note that the extra phase shift is a decreasing

Journal of Physics: Conference Series **359** (2012) 012013 doi:10.1088/1742-6596/359/1/012013

function of β/ω, and that the original KK relation of Eq. (6) is recovered when $\beta/\omega \to \infty$ [16]. Detailed theoretical considerations on the extra phase shift in various situations have been reported [17, 18]. In the case of actual experimental studies, however, the precise value of β may not be known. Accordingly, the value of β has been estimated from experimental $R_d(\omega)$ data by use of a combination of DL fitting and the modified KK transform [3, 8, 10, 13]. In this method, one uses Eq. (7) with $r(\omega) = \sqrt{R_d(\omega)}$ and looks for a value of β that well reproduces the $\sigma(\omega)$ given by a DL fitting of $R_d(\omega)$.

Note that, on the other hand, if the frequency range of interest is lower than the value of β, effects of the extra phase may be only minor, and the usual KK transform of Eq. (6), combined with the use of n_0=2.4 in Eqs. (3) and (4), might give sufficiently accurate values of optical constants. We will examine the validity of such a procedure in the next section.

3. Simulation with measured reflectance spectra

Here, we use $R_0(\omega)$ spectra actually measured on $PrRu_4P_{12}$ [19]. This compound is well known for showing a metal-to-insulator transition at about 60 K [20], and a clear energy gap in $\sigma(\omega)$ was observed in our previous work [19]. Here we use $R_0(\omega)$ at 60 K (metal) and 9 K (insulator) measured over a wide photon energy range of 0.008-30 eV [19], which are shown by the blue curve in Figs. 1(a) and 2(a). The procedure is the following.

(i) The full $R_0(\omega)$ spectrum is KK analyzed with Eq. (6) to obtain $n(\omega)$, $k(\omega)$ and $\sigma(\omega)$.

(ii) The above $n(\omega)$ and $k(\omega)$ are substituted into Eq. (1) with n_0=2.4 to derive $R_d(\omega)$ that is *expected* in a DAC.

(iii) The $R_d(\omega)$ obtained above is used with the usual KK transform of Eq. (6) and n_0=2.4 in Eqs. (3) and (4), to obtain $n(\omega)$, $k(\omega)$ and $\sigma(\omega)$. Before this is done, an appropriate cutoff and extrapolation are made to the $R_d(\omega)$, as described in detail below.

If the KK analysis on $R_d(\omega)$ works properly, the resulting $\sigma(\omega)$ from (iii) should well agree with that given by $R_0(\omega)$ and the usual KK analysis.

We first examine the 60 K data. The expected $R_d(\omega)$ at 60 K obtained by (i) and (ii) is indicated by the black curve in Fig. 1(a). In carrying out the integration in Eq. (6), the $R_0(\omega)$ spectrum were extrapolated below 0.008 eV and above 30 eV with the Hagen-Rubens and ω^{-4} functions, respectively. It is seen in Fig. 1(a) that $R_d(\omega)$ shows very high values above about 4 eV. This physically unrealistic feature resulted from the unphysical assumption of constant n_0=2.4 in the entire spectral range. In reality, of course, the refractive index of diamond cannot be constant and real near and above the band gap, where it shows strong light absorption. In addition, when $\omega \to \infty$, $n(\omega) \to 1$ and $k(\omega) \to 0$, and therefore $R_d(\omega) \to 0$ must hold just like any other material. Accordingly, before performing KK transform in (iii), $R_d(\omega)$ in Fig. 1(a) was cut off at some energy ω_c, and then it was extrapolated with ω^{-4} function. Several different values of ω_c were tried, as shown in Fig. 1(a). For each value of ω_c, the usual KK transform of Eq. (6) was made to get $\theta(\omega)$, which was then used to derive $n(\omega)$ and $k(\omega)$ with n_0=2.4 in Eqs. (3) and (4), and to finally obtain $\sigma(\omega)$.

Figure 1(b) shows the $\sigma(\omega)$ spectra obtained as described above, with different values of ω_c which are also indicated in Fig. 1(a). It is seen that the obtained $\sigma(\omega)$ spectra strongly depend on ω_c. With ω_c=2.0 eV, the resulting $\sigma(\omega)$ below 1.5 eV agrees very well with that derived from the original $R_0(\omega)$. (Actually, ω_c=2.2 eV gives the best agreement, but ω_c=2.0 eV data is shown instead. This is because the 2.2 eV data almost completely overlaps with that from $R_0(\omega)$, making it difficult to distinguish them in the figure.) The result for the 9 K data, where the sample is an insulator (semiconductor), is also shown in Fig. 2. A good agreement is again observed between the $\sigma(\omega)$ derived from the full $R_0(\omega)$ and that from $R_d(\omega)$ with ω_c=2 eV. The spectral range of their good agreement is below 1.5 eV, which is similar to the case of 60 K data discussed above. Note also that both 9 K and 60 K data show good agreement with the

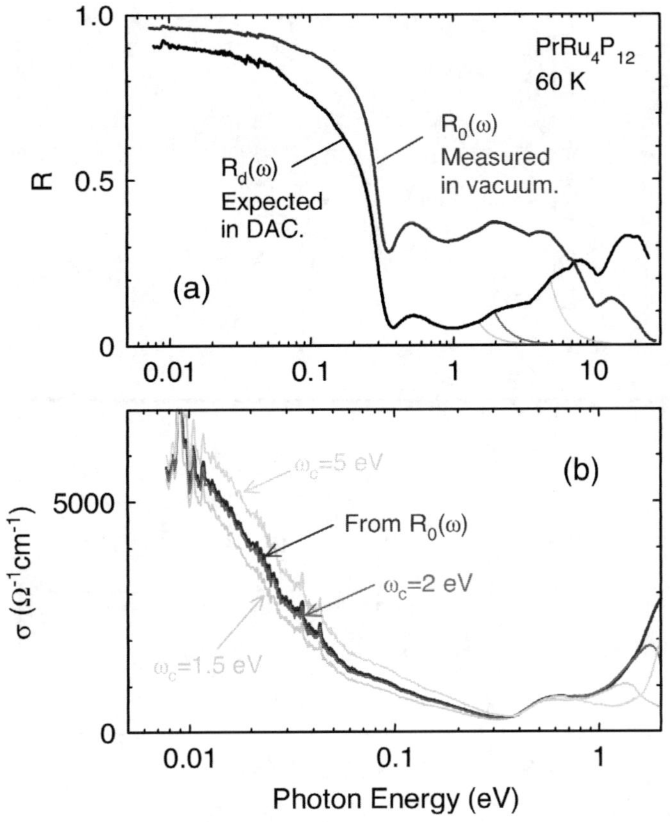

Figure 1. (a) $R_0(\omega)$ is the reflectance spectrum of $PrRu_4P_{12}$ measured at 60 K in vacuum [19], and $R_d(\omega)$ is that expected in a DAC calculated from $R_0(\omega)$ as described in the text. The green, red, and light blue curves are ω^{-4} extrapolations with cutoff energies of ω_c=1.5, 2, and 5 eV, respectively. (b) The optical conductivity (σ) obtained with KK analysis of $R_0(\omega)$ is compared with those obtained with KK analysis of $R_d(\omega)$ with a cutoffat ω_c=1.5, 2, and 5 eV, and ω^{-4} extrapolations above them. Below 1.5 eV, $\sigma(\omega)$ obtained from $R_d(\omega)$ with ω_c=2 eV agrees very well with that obtained from $R_0(\omega)$.

common value of ω_c= 2 eV. These results show that, for any spectral change in $R_d(\omega)$ (either temperature- or pressure-induced) below 1.5 eV, the corresponding $\sigma(\omega)$ can be obtained by the present method. In actual high pressure studies of strongly correlated materials with DAC [2-13], $R_d(\omega)$ is usually measured below 1-2 eV. Hence, above the high energy limit of $R_d(\omega)$ measurement, the $R_d(\omega)$ expected from $R_0(\omega)$ can be connected to the measured $R_d(\omega)$, with the cutoff and extrapolation discussed above. Then the connected $R_d(\omega)$ may be KK transformed to obtain $\sigma(\omega)$, as discussed above. An obvious condition required for this method to work properly is that the pressure- and temperature-induced changes of $R_d(\omega)$ should be limited below certain energy, which is 1.5 eV for $PrRu_4P_{12}$ as seen in Figs. 1(b) and 2(b). This condition is actually met in the high pressure data of $PrRu_4P_{12}$, which has enabled us to derive its $\sigma(\omega)$ under pressure up to 14 GPa using the present method [21].

We have also done similar simulations for other compounds, both metals and insulators, using actually measured data, and have obtained similar results. Namely, when an appropriate cutoff and extrapolation are applied to $R_d(\omega)$, the usual KK transform of Eq. (6) gave $\sigma(\omega)$

Journal of Physics: Conference Series **359** (2012) 012013 doi:10.1088/1742-6596/359/1/012013

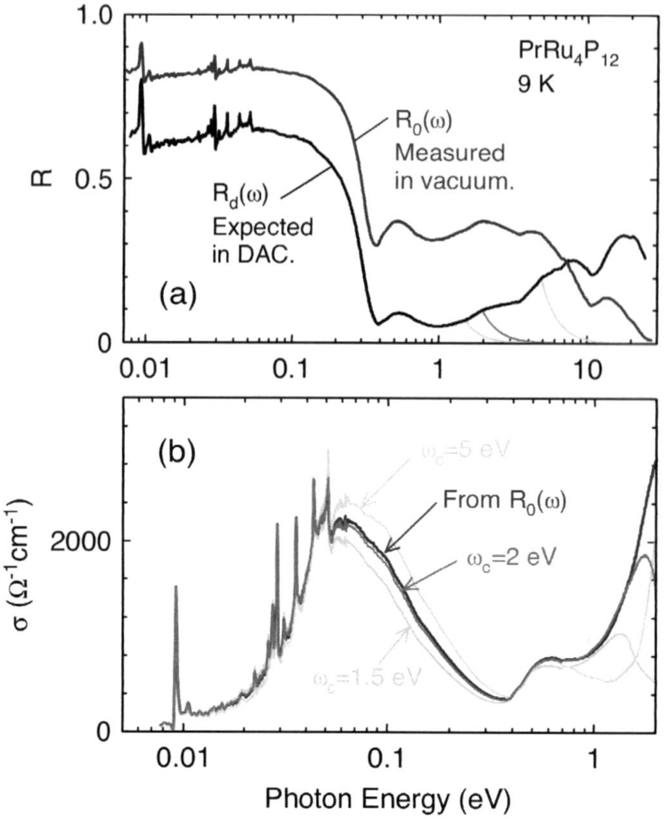

Figure 2. (a) $R_0(\omega)$ is the reflectance spectrum of $PrRu_4P_{12}$ measured at 9 K in vacuum [19], and $R_d(\omega)$ is that expected in a DAC calculated from $R_0(\omega)$ as described in the text. The green, red, and light blue curves are ω^{-4} extrapolations with cutoff energies of ω_c=1.5, 2, and 5 eV, respectively. (b) The optical conductivity (σ) obtained with KK analysis of $R_0(\omega)$ is compared with those obtained with KK analysis of $R_d(\omega)$ with a cutoffat ω_c=1.5, 2, and 5 eV, and ω^{-4} extrapolations above them.

spectra which agreed very well with those directly obtained from the wide range $R_0(\omega)$. A limitation of the present method is, as already mentioned above, it can give correct $\sigma(\omega)$ only below certain photon energy (1.5 eV in the case of $PrRu_4P_{12}$). Hence this method is useful when the temperature and pressure dependences of $R_d(\omega)$ is limited to below certain energy. In addition, to use the present method, it is required that $R_0(\omega)$ is known over a wide enough photon energy range, since $n(\omega)$ and $k(\omega)$ must be obtained from $R_0(\omega)$ with the usual KK analysis. While a mathematically rigorous justification of the proposed method is beyond the scope of this work, this method may be very useful as a simple analysis technique of reflectance spectra measured under high pressure with DAC.

Acknowledgments
This work has been done as a part of high pressure infrared studies of strongly correlated electron materials using synchrotron radiation at SPring-8, under the approval by JASRI (2009A0089 through 2011B0089). The $R_0(\omega)$ data used in Figs. 1 and 2 have been already published [19], in collaboration with M. Matsunami, L. Chen, M. Takimoto, T. Nanba, C. Sekine, and I. Shirotani.

References

[1] Basov DN, Averitt RD, van der Marel D, Dressel M 2011 *Rev. Mod. Phys.* **83** 471
[2] Syassen K, Winsen H, Zimmer HG, Tups H and Leger JM 1985 *Phys. Rev. B* **32** 8246
[3] Klehe A-K, McDonald RD, Goncharov AF, Struzhkin VV, Mao HK, Hemley RJ, Sasaki T, Hayers W and Singleton J 2000 *J. Phys.: Condens. Matter* **12** L247.
[4] Okamura H, Senoo K, Matsunami M, Nanba T 2007 *J. Phys. Soc. Jpn.* **76** Suppl. A, pp9.
[5] Matsunami M, Okamura H, Ochiai A, and Nanba T 2009. *Phys. Rev. Lett.* **103** 237202.
[6] Okamura H, Matsunami M, Kitamura R, Ishida S, Ochiai A and T. Nanba 2010 *J. Phys. Conf. Ser.* **125** 012051.
[7] Okamura H, Kitamura R, Matsunami M, Sugawara H, Harima H, Sato H, Moriwaki T, Ikemoto Y and Nanba T 2011 *J. Phys. Soc. Jpn.* **80** 084718.
[8] Pashkin A, Dressel M and Kuntscher CA 2006 *Phys. Rev.* B **74** 165118.
[9] Sacchetti A, Arcangeletti E, Perucchi A, Baldassarre L, Postorino P, Lupi S, Ru N, Fisher IR and Degiorgi L 2007 *Phys. Rev. Lett.* **98** 026401.
[10] Lavagnini M, Sacchetti A, Degiorgi L, Arcangeletti E, Baldassarre L, Postorino P, Lupi S, Perucchi A, Shin KY and Risher IR 2008 *Phys. Rev.* B **77** 165132.
[11] Mizuno T, Iizuka T, Kimura S, Matsubayashi K, Imura K, Suzuki H and Sato N 2008 *J. Phys. Soc. Jpn.* **77** 113704.
[12] Haumont R, Bouvier P, Pashkin A, Rabia K, Frank S, Dkhill B, Crichton WA, Kuntscher CA and Kreisel J 2009 *Phys. Rev. B* **79** 184110.
[13] Lupi S, Baldassarre L, Mansart B, Perucchi A, Barinov A, Dudin P, Papalazarou E, Rodolakis F, et al. 2010 *Nat. Commun.* **1** 105.
[14] Wooten F 1972 *Optical Properties of Solids* (Academic Press, New York)
[15] Dressel M and Gruner G 2002 *Electrodynamics of Solids* (Cambridge University Press, England)
[16] Plaskett JS and Schatz PN 1963 *J. Chem. Phys.* **38** 612.
[17] Bardwell JA and Dignam MJ 1985 *J. Chem. Phys.* **83** 5468.
[18] Bertie JE and Lan Z 1996 *J. Chem. Phys.* **105** 8502.
[19] Matsunami M, Chen L, Takimoto M, Okamura H, Nanba T, Sekine C and Shirotani I 2005 *Phys. Rev. B* **72** 073105
[20] Sekine C, Uchiuni T, Shirotani I and Yagi T 1997 *Phys. Rev. Lett.* **79** 3218
[21] Okamura H 2011, unpublished work.

6th Workshop on Infrared Spectroscopy and Microscopy with Accelerator-Based Sources IOP Publishing
Journal of Physics: Conference Series **359** (2012) 012014 doi:10.1088/1742-6596/359/1/012014

Infrared Microspectroscopy study of insulin crystals at high pressure

F. Piccirilli[a,c], S. Mangialardo[b], S. Lupi[b,c], P. Postorino[b], A. Perucchi[c]

[a] Dipartimento di Fisica, Università di Trieste, Italy.
[b] Dipartimento di Fisica, Università di Roma, "Sapienza", Italy.
[c] Elettra Sincrotrone Trieste, Italy.

Email: federica.piccirilli@libero.it

Abstract. During the last years the coupling of high pressure techniques and infrared spectroscopy has proven to be a very powerful tool in the study of conformational changes of proteins. Protein unfolding and monomerization are events that are expected to take place at high pressure due to the peculiarity of pressure to shift the system towards the state that occupies the minimum volume. We observed the growth of apparently cubic crystals at a pressure of about 4 kbar, subjecting to high pressure a solution of misfolded insulin. Even if high pressure is commonly used to tune the growth rate of crystals, protein crystallization at high pressure is not a well known process and no evidences of the particular case of insulin are present in literature.

1 Introduction

The amyloid diseases, which include Alzheimer's and Parkinson's diseases, the spongiform encephalopathies and type II diabetes, are characterized by the self-assembly and deposition of proteic material into insoluble ordered aggregates. So far, nearly 20 different amyloidogenic precursors have been identified although the detailed mechanism that leads to the formation of fibrils by these proteins is still unclear at a molecular level. For the difficulty in obtaining well diffracting crystals the atomic structure of fibrils has not yet been resolved. The accepted structural model for fibrils predicts the existence of a core protofilament composed of elements of β-sheets[1]. The sheets run parallel to the axis of the fibril and their component β-strands is thought to be perpendicular to this axis. The fibril is considered mature when the lateral packing between few extended protofilaments occurs. Insulin is a 51 residues protein central to regulate carbohydrate and fat metabolism in the body. It shows a high tendency to aggregate into amyloid-like fibrils *in vitro*, during its storage as pharmaceutical in the treatment of some forms of diabetes[2]. Due to the clinical relevance of insulin misfolding and its high tendency to aggregate into amyloid fibrils it is considered a model protein in the study of misfolding. When insulin is heated at high temperature and low pH, a series of structural changes occurs and results in the formation of the ordered aggregates[3]. High Hydrostatic Pressure has been proven to be a very powerful tool in the study of misfolding due to its effect on the energetic of the system that involves only variation on the volume contribution of Gibbs free energy[4]. High Pressure perturbations only affect the H-bonds of the system, tuning the conformation of the molecule towards the state that occupies the smallest volume. It causes the electrostriction of charged and polar groups, the elimination of packing defects, and the solvation of hydrophobic groups. Cavities and packing defects are expected to be major contributors to volume changes and their presence will make the system more susceptible to pressure unfolding/dissociation. In the last years many studies remarked the high pressure ability to monomerize physiological assemblies and to unfold native. Moreover, the high pressure induced rupture of amyloid fibrils has been recently reported[5]. Since amyloids are temperature insensitive, pressure represents a unique tool for studying their thermodinamical stability. Infrared spectroscopy is a well established method for the analysis of protein secondary structure. The singular advantage of FTIR over other techniques is that spectra can be obtained for biological samples in a wide range of environments, and proteins can be studied in their physiological conditions. There is a wealth of information that can be used to derive structural information by analyzing the shape and position of bands in the Amide I region (1600-1700 cm^{-1}) of the spectrum. The presence of a number of Amide I band frequencies have been correlated with the presence of α-helical, antiparallel and parallel β-sheets and random coil structures[6]. In particular, it is now well accepted that the occurrence of fibrillation can be easily followed on infrared spectra from the observation of a very narrow component at about 1620 cm^{-1}.

We studied the high pressure behavior of insulin fibrils in solution in the pressure range between 2.8 kbar and 11.4 kbar. During the compression cycle, we observed the formation of apparently cubic crystals. We present here the infrared microspectroscopy (IRMS) characterization of the high pressure crystals obtained. To determine their secondary structure, the crystals have been explored through the use of synchrotron light that allowed measurements on single crystals.

2 Experimental technique

Sample Preparation
Insulin fibrils have been prepared following the head-induced BPI fibrillation protocol[3]. We dissolved 11mg of BPI (Bovine Pancreatic Insulin) powder in 1ml of D_2O at acidic pD (~1.5) (the pD has been adjusted adding DCl to the solution). The protein solution has been incubated at 70°C for 6 hours, then loaded into the diamond anvil cell.

HP-FTIR measurements
The high pressure measurement has been performed through the use of a diamond anvil cell equipped with a screw mechanism to tune pressure. We used IIa diamonds with a culet diameter of 600 μm, and a 40 μm stainless steel 316L gasket, with a hole diameter of 200 μm. The pressure range explored goes from 2.8 kbar to 11.4 kbar.
The absorption measurement has been done with a Hyperion IR microscope in the frequency range between 1000 and 4000 cm^{-1} , at a resolution of 4 cm^{-1} and 256 scans, using a Vertex Bruker interferometer coupled with the conventional Globar source and the MCT detector. The spatial distribution of the protein inside the pressure cell has been determined through the use of the FPA detector.
The single crystal absorption measurement has been performed coupling the infrared interferometer with the synchrotron source, the infrared microscope and the MCT detector. The beam size for synchrotron measurements on single crystals has been set to 10x10μm, to fit the dimension of crystals. We loaded into the cell, together with the sample, a small ruby sphere as pressure calibrant. Pressure calibration has been performed through the ruby fluorescence technique, using an home-made fluorescence system mounted on the infrared microscope.

3 Results

BPI (Bovine Pancreatic Insulin) fibrils have been subjected to a compression cycle, exploring a pressure range between 2.8 kbar and 11.4 kbar. In Figure 1 is reported the Amide I band of the whole sample during the compression cycle. Together with the high pressure spectra, of the sample loaded into the diamond anvil cell, we reported the ambient pressure measurement, done using a standard liquid cell. The ambient pressure spectrum is composed of two β-extended peaks, at 1619 and 1630 cm^{-1} and other components absorptions coming from intramolecular structures: 1647 cm^{-1} (random coil), 1657 cm^{-1} (α-helix), 1663 cm^{-1} (3^{10}-helix or sheets), 1678 cm^{-1} (turns, loops and others). The presence of two peaks associated to the extended structure is due to the split of the lower frequency peak (1619 cm^{-1}) into two components. This phenomenon takes place when lateral packing between protofilaments occurs. We observed a very huge effect on the protein structure even at the lowest pressure applied (2.8 kbar). The components that can be found from the observation of the high pressure spectra are the following: 1619 cm^{-1} (β-extended), 1636 cm^{-1}(intramolecular β-sheet) , 1652 cm^{-1} (α-helix), 1672 cm^{-1} and 1684 cm^{-1}(turns, loops and others).

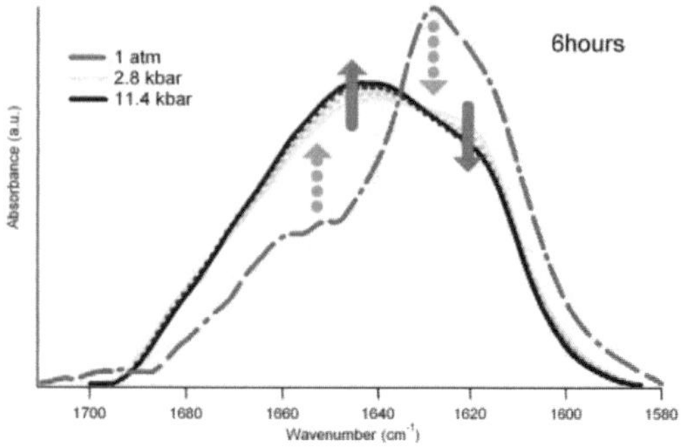

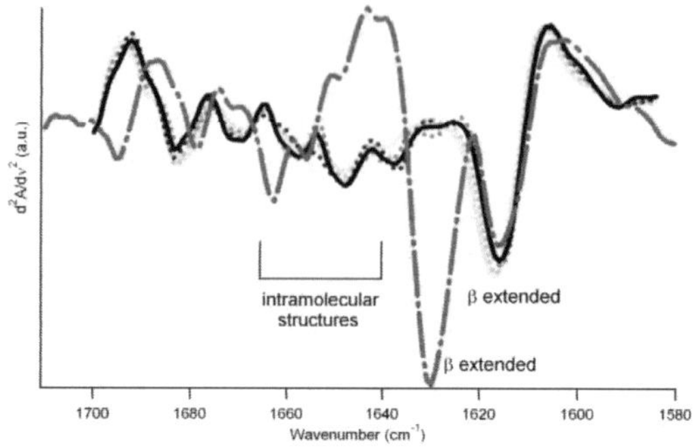

Figure 1: Absorbance (left) and 2nd-derivative (right) spectra of BPI fibrils at ambient pressure and during the compression cycle.

The presence of only one β-extended component at high pressure, remarked by the 2nd-derivative spectra, hints the absence of lateral packing between protofilaments, the solution is thus mainly populated by protofibrils. The Amide I area in the region of β-extended structures (1600-1640 cm^{-1}) significatively decreased during compression, sign of the induced rupture of part of the H-bonds composing the protofilaments. Conversely, the Amide I area above 1640 cm^{-1}, where random coils and helices absorb, increased.

At a pressure of about 4 kbar, after about 1 hour from the pressure tuning, a small crystal appeared.

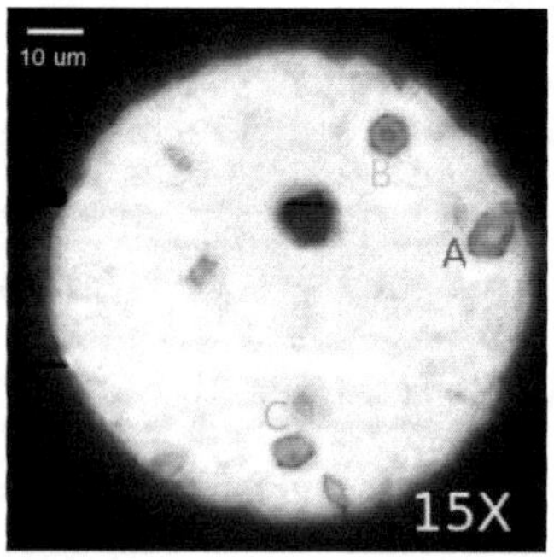

Figure 2: Picture of the sample after 4 days from the appearance of the first crystal (C). It is possible to recognize the pressure calibrant (black spot) in the upper part of the picture and ten small crystals all around it.

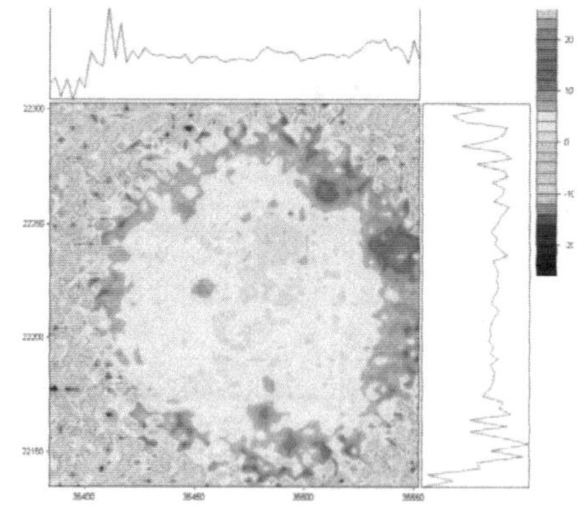

Figure 3: Amide I integrated signal of the sample after 4 days from the appearance of the first crystal. The image as been acquired through the use of the FPA detector.

The pressure has been kept constant to this value and after 4 days, about 10 crystals grew inside the cell (Figure 2). Crystals structure seemed to be apparently cubic. We acquired the FPA map of the whole sample surface to verify the spatial distribution of the protein inside the cell. The intensity distribution of the Amide I band inside the cell confirmed that the crystals were made of proteic matter (Figure 3).

To better characterize the secondary structure composition of crystals we also acquired single crystal absorbance (Figure 4). In Figure.5 are reported the Amide I bands of the three biggest crystals in the sample and of the soluble insulin acquired at 10x10 μm.

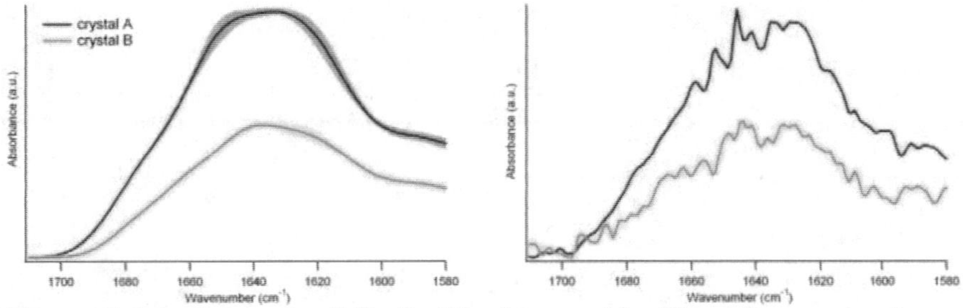

Figure.4 : Single crystal absorption spectra. (left) after 25-points smoothing filter and averaged on 5 measurements. The standard deviation is reported as shadow. (rhigt) original spectra.

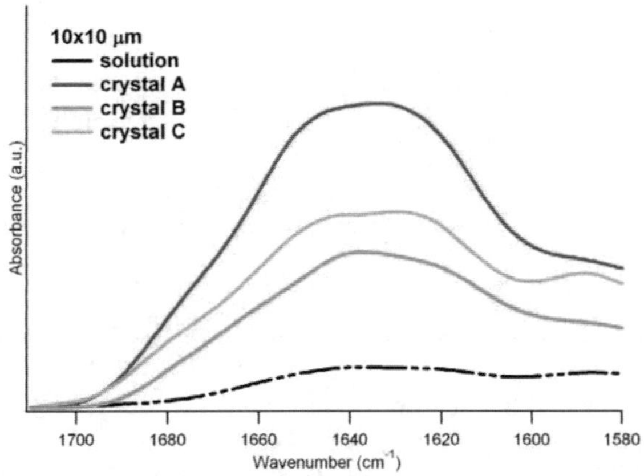

Figure.5 : Amide I band of the three biggest crystals inside the DAC. In dashed line is reported the absorbance of the solution around the crystals. Measurement performed with the synchrotron source and a beam size of 10x10μm.

For a more accurate evaluation of the secondary structure of the protein in crystalline form, it was necessary to consider also the contribution to the absorption coming from the optical path length occupied by the liquid sample. We approximated the shape of crystals to perfect cubes. To evaluate the true path length occupied by the soluble protein, we considered the difference between the thickness of the gasket, of 40 μm, previously calculated through the analysis of interference fringes, and the approximate height of the crystal, 10 μm, deduced from the observed lateral size of crystals. According to all these considerations, 3/4 of the total path length is occupied by the liquid sample, we thus subtracted the spectra following the equation:

$$Abs_{crystal} = 4 \times \left(Abs_{crystal-exp} - \frac{3}{4} Abs_{sol-exp}\right)$$

Where $Abs_{crystal\text{-}exp}$ and $Abs_{sol\text{-}exp}$ are the signals acquired respectively on single crystal and on the solution at 10x10 μm, shown in Figure 5.
In Figure 6 are reported the single crystal Amide I bands after the subtraction procedure.

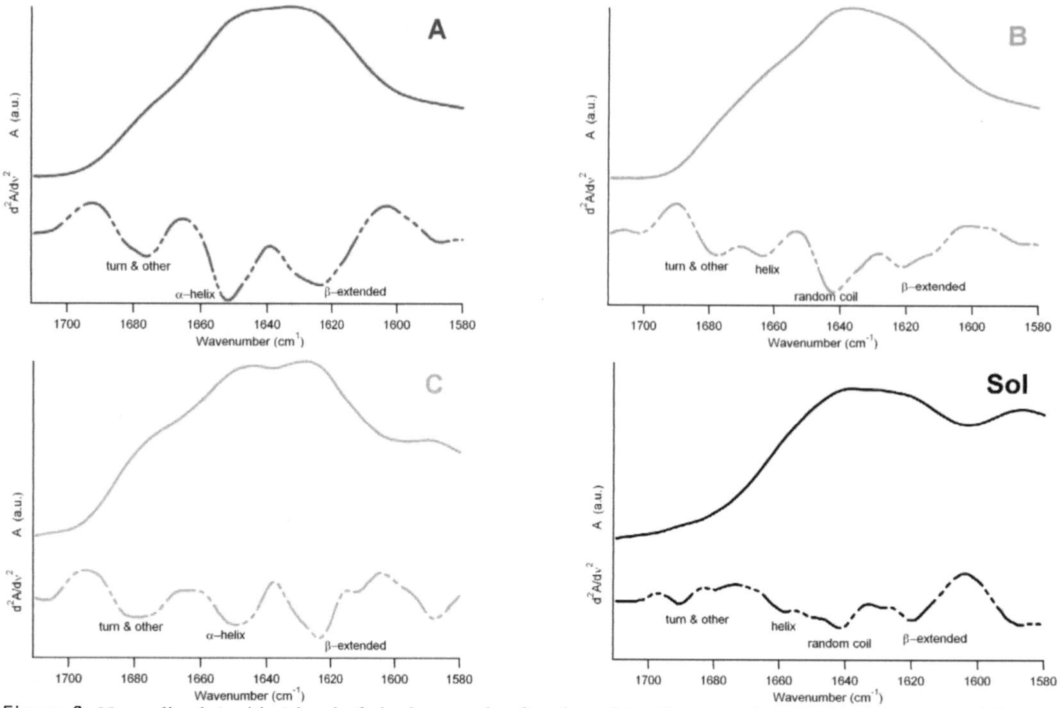

Figure 6: Normalized Amide I band of single crystals after the subtraction procedure. In black is reported the normalized Amide I band of the subtracted solution.The measurements have been performed with synchrotron beam size of 10x10 μm.

The subtracted single crystal bands are peaked around 1640 cm^{-1}, with three shoulders around 1678 cm^{-1}, 1646 cm^{-1} and 1622 cm^{-1}. From the 2nd-derivative we found the presence of β-extended structures in all the three crystals measured, suggested by the minima occurring around 1622 cm^{-1} (Figure.6). Except for the crystal B, that possesses a component around 1649 cm^{-1}, corresponding to random coils absorption, the presence of a significant amount of α-helices in the A and B crystals is evident by the huge minima around 1656 cm^{-1}.
The secondary structure of the soluble protein, evaluated by the observation of the spectrum of the solution at 10x10 μm, consists of all the absorptions found for crystals, with a very pronounced random coil component that makes it more similar to the crystal B.
We can therefore reasonably conclude that the crystalline form we obtained is composed of a partially folded intermediate.
The crystals have been compressed to 11 kbar then decompressed to 4 kbar. We found that they are stable in all the explored pressure range. After the pressure release the crystals slowly (about 4 minutes) disappeared. It is worth to notice that dimeric and hexameric insulin crystallizes at ambient pressure into cubic structures of similar dimensions[7].

4 Conclusions

The stabilization of insulin crystals under high pressure has been observed starting from a solution of fibrillated insulin and their secondary structure have been studied through the use of synchrotron radiation in the mid infrared range.
Based on the experimental findings, the following conclusions can be drawn:

1. A pressure of about 2.8 kbar can induce the rupture of the lateral packing between insulin amyloid-like protofilaments and the fragmentation of the fibrillar elongated structure.

2. A partially folded intermediate of insulin crystallizes at about 4 kbar and remains stable up to 11 kbar. The crystals have a size of about 10 μm and are apparently cubic.

The question that arises from the present work concerns the hypothetical full reversibility of the insulin fibrillation process , at higher pressures, with the possible occurrence of the native protein refolding. Moreover, our experimental findings leave room for the eventual X-ray characterization of the crystallized structures. This could help to confirm data interpretation and eventually clarify the still unresolved atomic structure of amyloid-fibrils.

References

[1] C.C. Lee, A. Nayak, A. Sethuraman, G. Belfort, G.J. McRae (2007). A Three-Stage Kinetic Model of Amyloid Fibrillation, *Biophysical journal.* 92(10), 3448-3458 .
[2] J. Brange (1987). Galenics of Insulin: the Physico-Chemical and Pharmaceutical Aspects of Insulin and Insulin Preparations, *Springer-Verlag*, Berlin.
[3] M. Bouchard, J. Zurdo, E.J. Nettleton, C.M. Dobson, and C. V. Robinson (2000). Formation of insulin amyloid fibrils followed by FTIR simultaneously with CD and electron microscopy, *Protein Sci.* 9, 1960–1967.
[4] K Heremans (1982). High Pressure Effects on Proteins and other Biomolecules, *Annual Review of Biophysics and Bioengineering* 11, 1-21.
[5] D. Foguel, M.C. Suarez, A.D. Ferrão-Gonzales, T.C. R. Porto, L.Palmieri, C.M. Einsiedler, L.R. Andrade, H.A. Lashuel, P.T. Lansbury, J.W. Kelly, and J.L. Silva (2003). Dissociation of amyloid fibrils of α-synuclein and transthyretin by pressure reveals their reversible nature and the formation of water-excluded cavities, *PNAS,* 100*(17),* 9831-9836.
[6] A. Barth and C. Zscherp (2002). What vibrations tell us about proteins. *Quarterly Reviews of Biophysics,* 35(4), 369-430.
[7] Badger, J., M. R. Harris, C. D. Reynolds, and A. C. Evans. (1991). Structure of pig insulin dimer in the cubic crystal. *Acta Crystallogr,* B47, 127–135.

Journal of Physics: Conference Series **359** (2012) 012015 doi:10.1088/1742-6596/359/1/012015

Combined far- and near-field chemical nanoscope at ANKA-IR2: applications and detection schemes

Diedrich A. Schmidt[1,2], Erik Bründermann[1], and Martina Havenith[1]

[1] Physical Chemistry II, Ruhr-Universität Bochum, 44780 Bochum, Germany

[2] Present address: Department of Physics, North Carolina A&T State University, Greensboro, NC 27411, USA

E-mail: Diedrich.Schmidt@rub.de, Erik.Bruendermann@rub.de, Martina.Havenith@rub.de

Abstract. A newly developed microscopy and nanoscopy station that combines far- and near-field microscopy with other microscopy modalities has recently been integrated at the ANKA-IR2 beamline. The various modalities include broadband synchrotron radiation and tunable laser-based near-field microscopy, atomic force microscopy, Raman microspectroscopy, and confocal laser and fluorescence microscopy. This multi-modal nanoscope is designed to combine a broad array of techniques to study the "same" sample at the "same" position. We show some examples that demonstrate several of the available modalities. We also discuss various detection schemes to facilitate sensitive absorption and reaction-kinetic experiments.

1. Introduction

A multi-modal nanoscope, which includes far- and near-field modalities, was recently integrated into the ANKA-IR2 (Karlsruhe, Germany) beamline in 2010. The far-field modalities include various confocal microscopies: optical, laser, Raman, and fluorescence. The near-field modality uses either aperture or aperture-less (scattering) methods in either the optical, infrared (IR), or terahertz (THz) range. By utilizing a metallic nano-finger, the infrared near-field modality can achieve chemical mapping with nanoscale spatial resolution using either currently available tunable laser-based sources or the high brilliance of the ANKA-IR2 beamline in the future. Laser-based infrared scattering-scanning near-field microscopy (s-SNOM) is a powerful chemical nanoscopy technique which has been useful in determining complex dielectric constants at the nanoscale [1], characterizing crystal structure and phonon-plasmon coupling [2, 3], and surface phonon polaritons [4]. Recently, broadband IR [5, 6] and THz [7] nanoscopy has also been shown.

Using laser-based infrared scattering SNOM, we have investigated self-assembled monolayers [8], nanoscale lipid membranes [9], nanografted patches of mixed single- and double-stranded DNA [10], ultra-thin mixed polymer brush films [11, 12], as well as "single" nanoparticles [13], and sub-surface [14] and activated [15] dopants in semiconductors. Utilizing the other modalities available at the ANKA-IR2 nanoscope, we have also investigated the status of mitochondria in UV-irradiated human spermatozoa [16], tracked drug-delivery compounds in live cells [17], and determined variations in surface charge densities of epitaxial graphene by performing a combined confocal Raman and atomic force microscopy study [18]. Here, we report on the status of the instrument at the ANKA-IR2 beamline [19]. We introduce the nanoscope experimental station and present recent examples illustrating the different modalities of the instrument, followed by a discussion on different detection schemes.

Journal of Physics: Conference Series **359** (2012) 012015 doi:10.1088/1742-6596/359/1/012015

2. Nanoscope station at ANKA

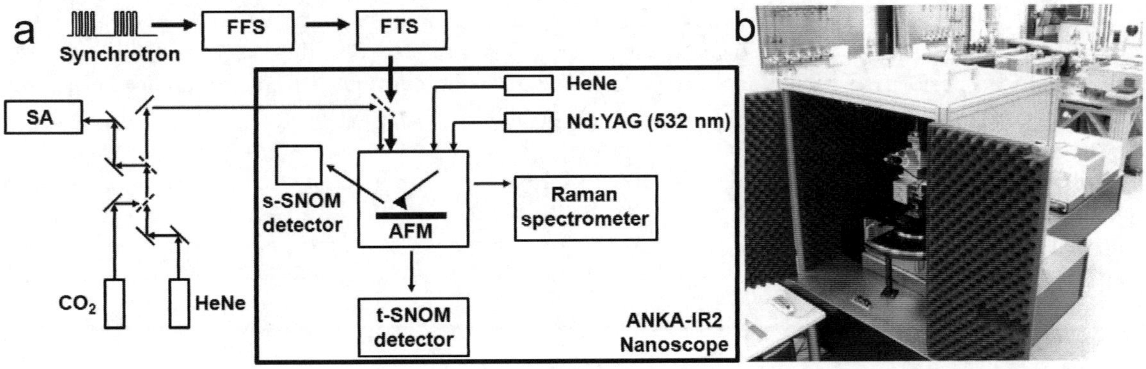

Figure 1. a) Schematic layout of the Nanoscope station at the ANKA-IR2 beamline. The synchrotron radiation passes through a fast feedback system for beam stabilization (FFS) before entering the Fourier transform spectrometer (FTS) and then being focused on a cantilever of an atomic force microscope (AFM). The external tunable CO_2 and alignment HeNe lasers, and the spectrum analyzer (SA) for the CO_2 laser, are also shown. Arrows represent beam paths. Solid lines represent beam steering optics while dashed lines are movable optics, e.g. flip mirrors, to choose between different radiation sources. The nanoscope also has two fiber-coupled "internal" lasers for Raman or SNOM, a HeNe and frequency-doubled Nd:YAG (532 nm) laser. The SNOM modality can be operated in either scattering mode (s) or transmission mode (t). The Raman and t-SNOM modalities use optical sources, while s-SNOM can use either synchrotron IR and THz, or optical sources. b) Photograph of nanoscope (foreground) in the experimental hutch of IR2. The FTS can be seen behind the enclosure box of the nanoscope. Photograph courtesy of B. Gasharova.

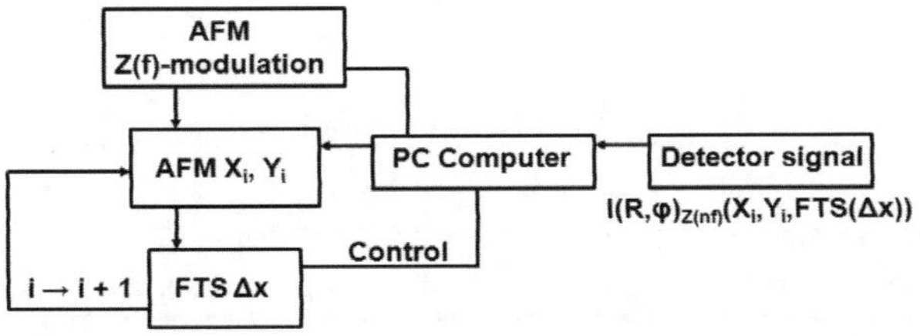

Figure 2. Flow diagram of the measurement scheme. The modulation of the AFM cantilever is fed to the computer (PC) and is used as feedback for the position of the sample. At each point (X_i, Y_i), a spectrum is acquired. The FTS performs either a continuous scan or a step-scan measurement at two different positions of the interferometer. The detector signal is demodulated at the nth harmonic, which is set by an external lock-in amplifier.

The nanoscope experimental station is coupled to the broadband synchrotron radiation of the ANKA-IR2 beamline via a fast feedback system for beam stabilization and a vacuum Fourier-transform spectrometer (FTS, Bruker Optik, model VERTEX 80v), with a far-infrared/THz spectral extension to $\sim$10 cm^{-1} (see Fig. 1). After the light passes through the Fourier transform spectrometer, it is coupled to the microscope using broadband mirror optics to access both the THz and IR radiation from the beamline. The microscope (WITec, model alpha300ASR) is already equipped for three measurement techniques: atomic force microscopy, scanning near-field optical microscopy, and confocal Raman microscopy. The fourth technique, infrared scattering

Journal of Physics: Conference Series **359** (2012) 012015 doi:10.1088/1742-6596/359/1/012015

SNOM, utilizes a metallic probe of an atomic force microscope (AFM) in combination with the IR radiation from the beamline or, when used off-line for pre-alignment or preparatory studies, a tunable CO_2 laser. The SNOM modality in transmission uses specialized aperture tips for top illumination, and can be performed with either a 532 nm or 633 nm laser, while scattering SNOM can be performed with a separate 633 nm laser. The confocal Raman microscopy (cRM) modality can be operated with either the 532 nm or 633 nm laser, with diffraction-limited lateral resolution that depends on the laser source. For scattering SNOM, we operate the nanoscope in a non-contact mode of the AFM in order to modulate the near-field signal and use higher-order harmonic demodulation [20–22] to eliminate background radiation and enhance the near-field contrast [23]. Figure 2 shows the flow diagram for the measurement scheme with a broadband radiation source.

3. Modalities

In the following, we present recent examples highlighting the optical and IR near-field (SNOM), far-field confocal Raman, and atomic force microscope (AFM) modalities of the ANKA-IR2 nanoscope.

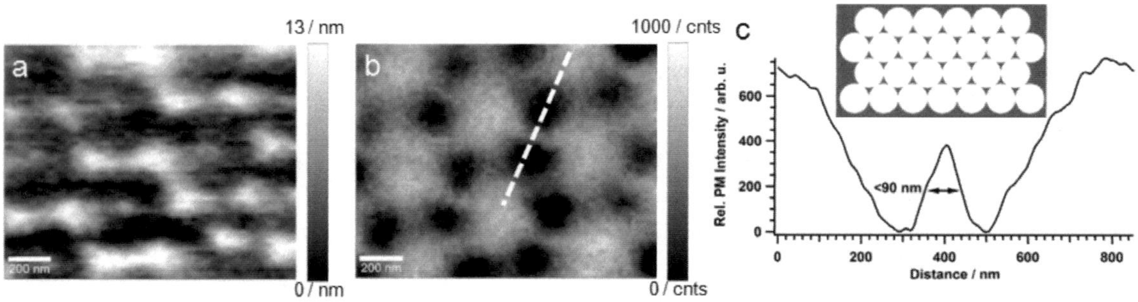

Figure 3. a) Topography of a metallic nano-structured test sample. b) Transmission SNOM image measured with a photomultiplier. c) line profile of the photomultiplier signal in b) along the dashed white line showing an optical resolution of <90 nm at $\lambda = 532$ nm, corresponding to $\sim \lambda/6$; the inset is a schematic of the nano-structured sample. Scale bars in a) and b) are 200 nm.

For near-field imaging in the optical regime, we use a metallic nano-structured test pattern made from silver. The light from the 532 nm laser is focused onto the apex of the specially designed nanocone. In transmission SNOM, we achieved simultaneous optical and spatial resolution of less than 90 nm. Figure 3a) shows the topography of the metallic nanostructures measured by the specialized aperture tips, while Fig. 3b) shows the simultaneously acquired transmission SNOM image recorded by a photomultiplier.

For scattering SNOM, we use a "soft" matter test sample of a microcontact-printed line pattern on a gold coated substrate consisting of double stranded DNA. Figure 4a) shows the topography of the DNA (bright vertical stripes in Fig. 4a)) surrounded by a self assembled monolayer of octadecanethiol (ODT), which fills the open space in-between the DNA stripes by binding to the gold surface of the substrate. The simultaneously acquired near-field image is shown in Fig. 4b). The wavelength of $\lambda = 9.5$ μm (from a CO_2 laser), coincides with an absorption maximum of DNA. The dark vertical stripes correspond to DNA, while ODT has no absorption bands near 9.5 μm, resulting in a clear contrast between the two materials. The subtle variations in intensity of the near-field image are due to interference with the background signal. More complex nanostructures can be made with the available nanolithography software using

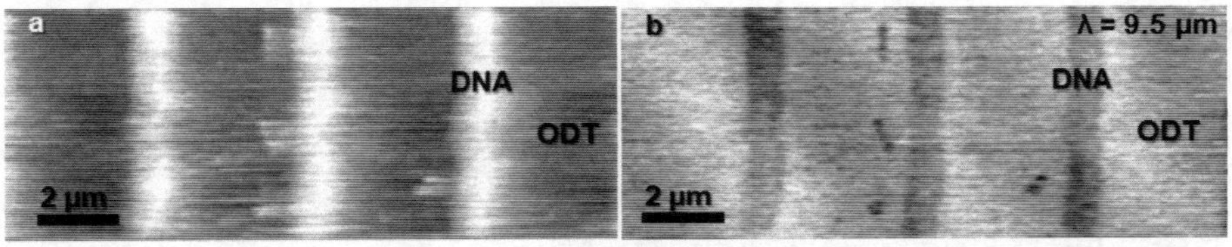

Figure 4. a) Topography of a microcontact-printed structured monolayer of double-stranded DNA separated by octadecanethiol (ODT); the DNA self-assembles in a monolayer which is ~20 nm higher (bright, narrow stripes) than the ODT (darker, wider stripes). b) Simultaneously measured near-field image of the sample at $\lambda = 9.5$ μm with a resolution of $< \lambda/100$, near the absorption maximum of DNA. The dark bands in the image correspond to the DNA stripes, while the ODT has no absorption near 9.5 μm. Scale bar is 2 μm in both images.

nanografting with the tip of an atomic force microscope cantilever in a liquid environment. For more details that demonstrate the complex nanografting of mixed, single- and double-stranded DNA patches, and their subsequent spectral read-out by scattering SNOM, we refer the reader to Ref. [10].

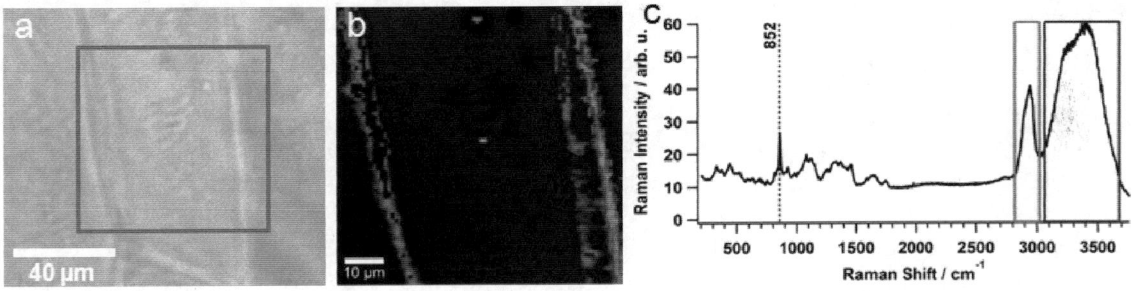

Figure 5. a) Optical image of an onion cell showing the scan area in red. b) False-color chemical map of the onion cell in a) showing the cell walls (red), primary water content (blue), and the amino acid tyrosine (green). c) Representative Raman spectrum showing the three primary bands used to create the image in b): tyrosine band at 852 cm^{-1}, C-H stretch corresponding to the cell walls (red box), and the O-H stretch of water (blue box). Scale bars in a) and b) are 40 μm and 10 μm, respectively.

One of the highly versatile modalities of the nanoscope is confocal Raman microscopy, as it offers a non-invasive, quantitative, and non-destructive method to examine living cells. In addition, it can be used to investigate liquid samples or samples containing a high water content. As an example, we looked at a fresh sample of onion cells prepared by peeling a single layer of onion skin and mounting it on a CaF_2 window. Figure 5a) shows the optical image of a fresh onion cell. Figure 5b) shows a false-color overlay image that highlights the large water content of the onion cell (blue), the cell walls (red), and the amino acid tyrosine (green). The individual chemical images used to create Fig. 5b) were made by integrating over the corresponding bands of interest and are shown in Fig. 5c): O-H stretch of water (blue box), C-H stretch which shows the structure of the cell walls (red box), and the amino acid tyrosine at 852 cm^{-1}. The ANKA-IR2 confocal Raman microscope included in the nanoscope system was also used to monitor the status of mitochondria in UV-irradiated human spermatozoa [16] and in tracking drug-delivery compounds [17] in living human cells. The system can be equipped with additional lasers in the future, such as 488 nm and 785 nm, to widen Raman applications in the analysis of living cells or tissue, semiconductor devices, or in materials analysis.

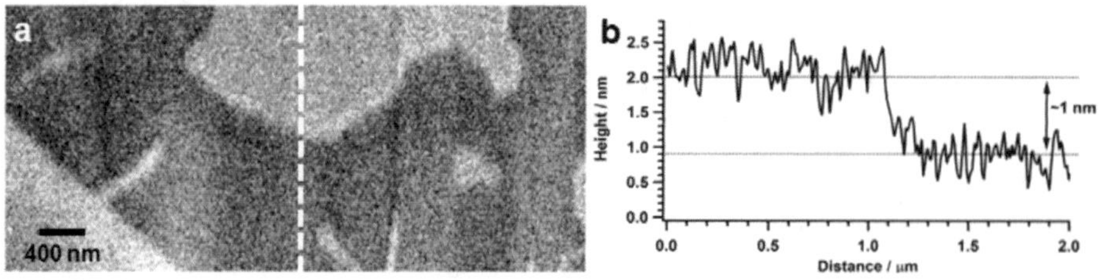

Figure 6. a) Atomic force microscope image of a partially graphitized SiC(0001) substrate. The bright region is SiC and the dark region is single layer graphene. b) Line scan along dotted line in a) showing the vertical resolution of ~1 nm. Scale bar in a) is 400 nm.

As a final example that illustrates the sensitivity of the atomic force micrscope modality, we studied a partially graphitized silicon carbide (SiC(0001)) sample. One of the growth modes of graphene on 6H-SiC(0001) is the conversion of three Si-C bilayers into a single sheet of graphene [24–26] by evaporation of the Si. Figure 6a) shows how the atomic force microscope resolves partially graphitized SiC(0001). The bright areas are SiC while the dark areas are single layer graphene. A line scan (dashed line in Fig. 6a)) across the growth front, with a step height of ~1 nm, is shown in Fig. 6b).

4. Detection schemes

In many applications, time-dependent processes at the nanoscale, such as dynamics in chemical reactions, kinetics at interfaces, energy level transitions in semiconductors and solid state materials, are of prime importance. For such applications, we study the pulse structure of the ANKA storage ring in the visible, IR, and THz spectral range [27] to be able to exploit the short pulses emitted by the storage ring. Figure 7 shows a measurement of a regular fill pattern during user operation. The 1 GHz avalanche photo diode used here resolves individual pulses that correspond to individual bunches (Fig. 7a,b). The structure, although not the intensity distribution (due to THz-field and electron bunch coherence effects), is similar to the THz pulse trains resolved with a hot electron bolometer [28]. The room temperature IR detector with a bandwidth of 200 MHz is not fast enough to resolve the individual pulses, but allows for the separation of the pulses originating from the three bunch trains in the ring (Fig. 7c). A room temperature IR detector (VIGO PVI-4TE-4, DoroTEK GmbH, Strausberg, Germany) with an increased bandwidth and risetime of ~0.5 ns partially resolves the separate IR pulses of each bunch (Fig. 7d). In the future, infrared hot electron bolometers could provide sufficient speed to resolve the 2 ns spaced single pulses in more detail.

Using the visible light, a trigger can be derived to detect the IR or THz radiation synchronous with the roundtrip time in the storage ring. One possibility to improve the signal-to-noise ratio in normal imaging modes is to use boxcar techniques to arm the detection branch only when the signal is present while removing excess noise during the time between pulses. The typical decay of the ring current with time or fluctuations from pulse to pulse can be compensated by a second, nearly identical detector monitoring the signal, or by using a balanced detection scheme. These methods can improve the image quality while at the same time learning more about the necessary conditions to obtain stable intensity and timing in storage rings.

5. Outlook

In a recent work [15], we have shown that it is possible to gain new insights in nanofabrication by studying the "same" sample at the "same" location, and using a multimodal approach with

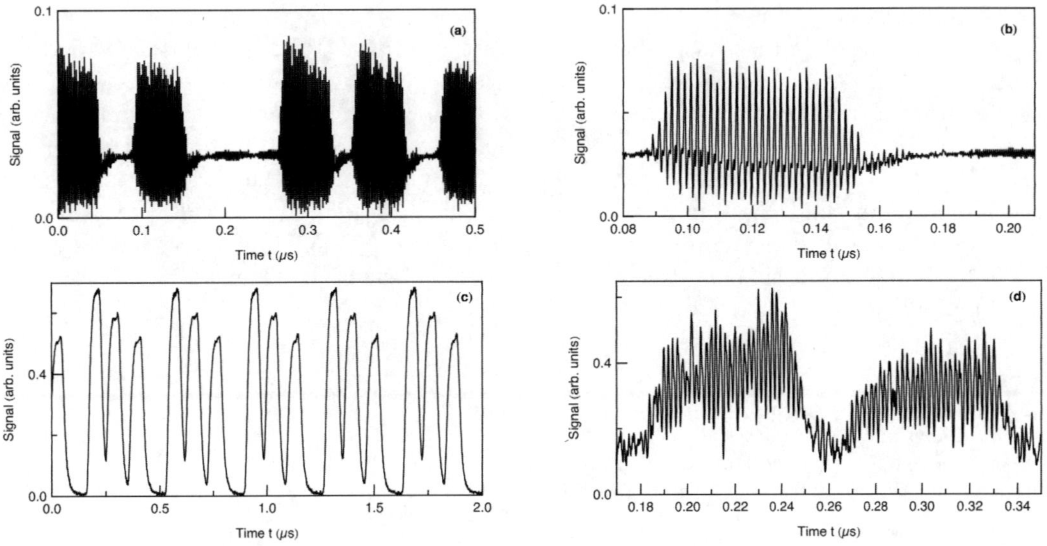

Figure 7. Measurements during normal X-ray user operation: (a) signal of ANKA pulses measured with avalanche photo diode for the visible spectral range, (b) zoom into trace of (a), (c) signal of ANKA pulses measured with a room temperature, 200 MHz bandwidth IR detector, (d) partially resolved pulses by an improved room temperature IR detector.

far-field synchrotron IR microscopy, confocal Raman microscopy, atomic force microscopy, and scattering near-field IR microscopy. The available multi-modal ANKA-IR2 nanoscope station provides the unique possibility of studying the same sample location using a "single" instrument with different modalities. The high brilliance and broadband synchrotron radiation of the ANKA-IR2 beamline, in combination with various laser sources, covers the optical, mid-IR, far-IR, and THz regions of the electromagnetic spectrum. The far-field confocal microscopy techniques are all diffraction limited, however the near-field methods approach resolutions of $\sim \lambda/1000$ [29], and are essentially limited by the size of the probe used for SNOM [30]. Improvements in spatial resolution of near-field imaging can be made by utilizing designer metallic waveguides [31, 32], which enhance the coupling and localize the electromagnetic field to smaller volumes.

6. Summary

In summary, we outlined the new nanoscope station that was recently installed at the ANKA-IR2 beamline. The multi-modal nanoscope, which incorporates far-field, near-field, and tactile methods, such as confocal microscopy (laser, optical, and Raman), SNOM in different modalities, and atomic force microscopy, offers the ability to gain new insights as the various methods can examine the "same" sample and "same" location without the need of moving to different instruments. We presented several examples of the possible measurement modalities of this instrument, including metal nano-structures for transmission SNOM, microcontact-printed DNA for scattering SNOM, fresh and wet onion cells for Raman microscopy, and partially graphitized SiC for atomic force microscopy. The emission pulse structure of the ANKA storage ring was examined and resolved in the visible and IR to investigate the possibility of measuring kinetics and dynamics, and we discussed different detection schemes for improving the signal-to-noise ratio and the imaging quality for future applications.

Acknowledgments
The project is developed in close collaboration with the ANKA-IR2 team, and we would like to especially thank B. Gasharova, Y.-L. Mathis, M. Süpfle, J. Steinmann, as well as A.-S. Müller and her team from the THz Radiation Group. The authors also appreciate the beam time made available at ANKA (Karlsruhe, Germany). M. H. thanks the BMBF for support under grants 05KS7PC2 and 05K10PCA. D. A. S. acknowledges support within the BMBF funded project. We thank I. Kopf for the DNA sample preparation and T. Ohta for the partially graphitized SiC sample. We also thank F. Ballout, M. Filimon, A. Bergner, G. Wollny, J.-S. Samson, and M. Mischo for valuable discussions and help, as well as K. Meister, E. Edengeiser, and A. Halang for handling cell cultures.

References

[1] Hillenbrand R and Keilmann F 2002 *Appl. Phys. Lett.* **80** 25
[2] Hillenbrand R and Keilmann F 2002 *Nature* **418** 159
[3] Huber A, Ocelic N, Taubner T and Hillenbrand R 2006 *Nano Letters* **4** 774
[4] Ocelic N and Hillenbrand R 2004 *Nature Materials* **3** 606
[5] Amarie S, Ganz T and Keilmann F 2009 *Optics Express* **17** 3430
[6] Huth F, Schnell M, Wittborn J, Ocelic N and Hillenbrand R 2011 *Nature Materials* **10** 352
[7] von Ribbeck H R, Brehm M, van der Weide D W, Winnerl S, Drachenko O, Helm M and Keilmann F 2008 *Optics Express* **16** 3430
[8] Kopf I, Samson J S, Wollny G, Grunwald C, Bründermann E and Havenith M 2007 *J. Phys. Chem. C* **111** 8166
[9] Wollny G, Bründermann E, Arsov Z, Quaroni L and Havenith M 2008 *Optics Express* **16** 7453
[10] Kopf I, Grunwald C, Bründermann E, Casalis L, Scoles G and Havenith M 2010 *J. Phys. Chem. C* **114** 1306
[11] Filimon M, Kopf I, Ballout F, Schmidt D A, Bründermann E, Rühe J, Santer S and Havenith M 2010 *Soft Matter* **6** 3764
[12] Filimon M, Kopf I, Schmidt D A, Bründermann E, Rühe J, Santer S and Havenith M 2011 *Phys. Chem. Chem. Phys.* **13** 11620
[13] Samson J S, Meißner R, Bründermann E, Böke M, Winter J and Havenith M 2009 *J. Appl. Phys.* **105** 064908
[14] Samson J S, Wollny G, Bründermann E, Bergner A, Hecker A, Schwaab G, Wieck A D and Havenith M 2006 *Phys. Chem. Chem. Phys.* **8** 753
[15] Ballout F, Samson J S, Schmidt D A, Bründermann E, Mathis Y L, Gasharova B, Wieck A D and Havenith M 2011 *J. Appl. Phys.* **110** 024307
[16] Meister K, Niesel J, Schatzschneider U, Metzler-Nolte N, Schmidt D A and Havenith M 2010 *Angewandte Chemie Int. Ed.* **49** 4540
[17] Meister K, Schmidt D A, Bründermann E and Havenith M 2010 *Analyst* **135** 1370
[18] Schmidt D A, Ohta T and Beechem T E 2011 *Phys. Rev. B accepted*
[19] Bründermann E, Schmidt D A, Kopf I and Havenith M 2010 *WIRMS 2009 5TH International Workshop on Infrared Microscopy and Spectroscopy with Accelerator Based Sources* vol 1214 p 7
[20] Ash E A and Nicholls G 1972 *Nature* **237** 510
[21] Knoll B and Keilmann F 2011 *Opt. Commun.* **182** 321
[22] Ocelic N, Huber A and Hillenbrand R 2006 *Appl. Phys. Lett.* **89** 101124
[23] Taubner T, Keilmann F and Hillenbrand R 2005 *J. Korean Phys. Soc.* **47** S213
[24] Borovikov V and Zangwill A 2009 *Phys. Rev. B* **80** 121406(R)
[25] Bolen M L, Harrison S E, Biedermann L B and Capano M A 2009 *Phys. Rev. B* **80** 115433
[26] Ohta T, Bartelt N C, Nie S, Thrmer K and Kellogg G L 2010 *Phys. Rev. B* **81** 121411(R)
[27] Müller A S, Birkel I, Casalbuoni S, Gasharova B, Huttel E, Mathis Y L, Moss D, Smale N, Wesolowski P, Bründermann E, Bückle T and Klein M 2008 *Proc. European Particle Accelerator Conference* vol WEPC046 (European Particle Accelerator Conference) p 2091
[28] Müller A S, Birkel I, Huttel E, Mathis Y L, Smale N, Bründermann E, Bückle T, Fitterer M, Hillenbrand S, Hiller N, Hofmann A, Judin V, Klein M, Marsching S, Sonnad K, Feikes J, Hartrott M, Hübers H W, Semenov A, Klein R, Müller R, Ulm G and Wüstefeld G 2009 *Proc. PAC09* p TU5RFP027
[29] Huber A J, Wittborn J and Hillenbrand R 2010 *Nanotechnology* **21** 235702
[30] Cvitkovic A, Ocelic N and Hillenbrand R 2007 *Optics Express* **15** 8550
[31] Berweger S, Atkin J M, Olmon R L and Raschke M B 2010 *The Journal of Physical Chemistry Letters* **1** 3427

[32] De Angelis F, Das G, Candeloro P, Patrini M, Galli M, Bek A, Lazzarino M, Maksymov I, Liberale C, Andreani L C and Fabrizio E D 2010 *Nature Nanotechnology* **5** 67

Journal of Physics: Conference Series **359** (2012) 012016 doi:10.1088/1742-6596/359/1/012016

Development of a Scanning Near-Field Sub-THz-Wave Microscopy with Coherent Transition Radiation

T Takahashi[1], T Iizuka[2], T Mori[3], and S Kimura[4, 2]

[1] Research Reactor Institute, Kyoto University, Osaka, JAPAN
[2] School of Physical Sciences, The Graduate University for Advanced Studies, Okazaki, JAPAN
[3] Graduate School of Pure and Applied Sciences, University of Tsukuba, Tsukuba JAPAN
[4] UVSOR Facility, Institute for Molecular Science, Okazaki, JAPAN

e-mail: takahasi@rri.kyoto-u.ac.jp

Abstract. In order to establish the technique of near-field microscopy using an accelerator-based coherent radiation, some basic properties are experimentally investigated with coherent transition radiation (CTR) from a linac. In the transmission mode, the spatial resolution power in accepting whole wavelengths of CTR was 4.4, which was obtained by scanning the metallic plate edge with a metallic conical cone as an illumination probe. It was found that the probe with a dielectric film effectively scattered the near-field. The dependence of the resolution power with the wavelength was investigated with a conical cone of Teflon coated by silver paste on the tapered surface as an illumination probe. The resolution power became large with increasing the wavelength. In the reflection mode, two types of optical configuration were prepared to detect the reflected light, i.e. the beam-splitter type and the light-guide type. The spatial resolution powers in the two configurations were 4.4 and 4.6, respectively.

1. Introduction

In recent years, various types of coherent radiation, i.e. coherent synchrotron radiation [1], transition radiation [2], Cherenkov radiation [3], diffraction radiation [4] and Smith-Purcell radiation[5], emitted from a relativistic electron beam have been investigated experimentally and attracted a considerable attention as a new and powerful light source in the THz-wave region. The technique of near-field THz-wave microscopy is a characteristic application of such brilliant coherent radiation. This technique provides high spatial resolution below the diffraction limit. In the storage ring BESSY-II (Germany), spatial resolution of $\lambda/40$ was obtained at 2 cm^{-1} using coherent synchrotron radiation [6]. In this paper some basic properties, i.e. spatial resolutions and spectra, have been experimentally investigated using coherent transition radiation (CTR) in order to establish the technique of near-field THz-wave microscopy with an accelerator-based coherent radiation.

2. Experimental Procedures

The experiment was performed at the coherent radiation beamline [7] at the 40-MeV L-band linac of the Research Reactor Institute, Kyoto University. The width of the macro pulse and the repetition rate of the electron beam were 47 ns and 46 Hz, respectively. The charge of a bunch was 1.2 nC. The

Journal of Physics: Conference Series **359** (2012) 012016 doi:10.1088/1742-6596/359/1/012016

schematic layout of this experiment is shown in Fig. 1. The sub-THz-wave source was the superposition of the forward CTR emitted from a titanium window 30 μm thick and the backward CTR from an aluminum-foil 15-μm thick. The CTR was guided to the experimental room as a parallel beam 15 cm in diameter and was detected by a liquid-helium-cooled Si bolometer through a spectrometer. This beamline is equipped with two interferometers and one monochromator. In this experiment the spectrum of CTR was measured by a Martin-Puplett type interferometer in a vacuum chamber as shown at the right-hand side in Fig. 1. The output signal from the detector was amplified by a lock-in amplifier synchronized with the repetition rate of the electron beam.

Aperture probes were used both in the transmission and the reflection mode as illumination and collection probes. In the reflection mode, two types of optical configuration were prepared in order to detect the reflected radiation, namely the beam-splitter type and the light-guide type. Their details are explained in section 3.2.

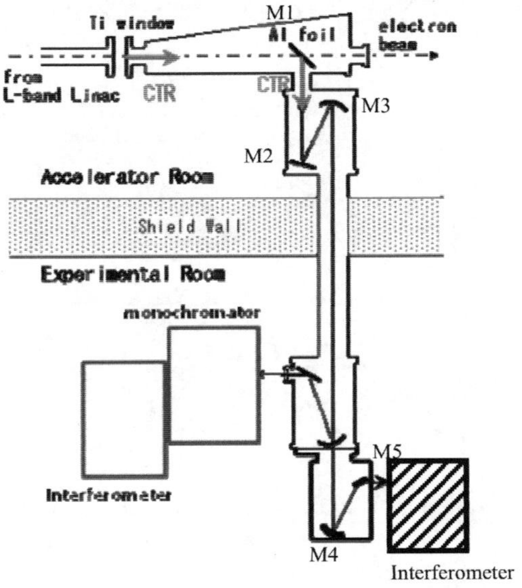

Figure 1. Schematic layout of the beamline at KURRI-LINAC. Keys are M1, a plane mirror of aluminium foil; M2, M5, plane mirrors; M3, M4, spherical mirros. The light source was the superposition of CTR from a titanium window and from an aluminium-foil mirror. A Martin-Puplett type interferometer in a vacuum chamber was used with a liquid-helium-cooled Si bolometer.

3. Results and discussion

3.1. Transmission mode

The metallic conical cone with an aperture 260 μm in diameter was used as an illumination probe. The detector was equipped with the collection probe with an aperture 3mm in diameter at the incident window. In order to investigate the spatial resolution, an opening of 1 mm in diameter on a stainless steel (SUS) sheet 50 μm thick was scanned in front of the aperture on the top of the illumination probe. The observed intensity of the transmitted radiation through the opening is plotted in Fig. 2 and its first derivative curve is shown by a dotted curve in Fig. 3. The solid curve of this figure is the Gaussian fitting. The width of the derivative curve is equivalent to the spatial resolution. Then, the spatial resolution was estimated to be 170 μm.

In Figure 4, the solid and dashed curves show the observed CTR spectra through the illumination probe and without it, respectively. The former intensity was multiplied by 1000 in this figure to clarify the difference of the spectral shape from the latter. The shift of the spectral distribution to shorter wavelengths arises from passing through the pinhole. Considering the wavelength of 750 μm at the peak intensity, the spatial resolution power was evaluated to be 4.4.

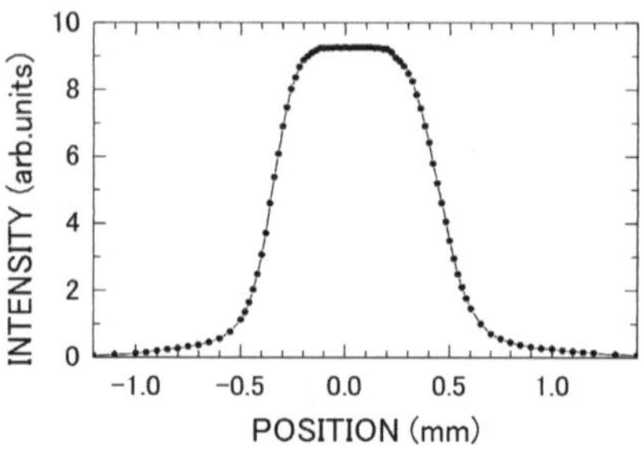

Figure 2. Scanning curve of an opening of 1 mm in diameter.

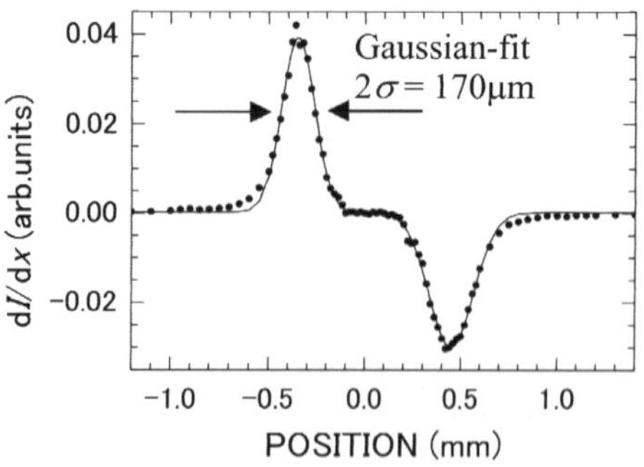

Figure 3. First derivative curve of Fig. 1 (dotted curve) and the Gaussian fitting (solid curve).

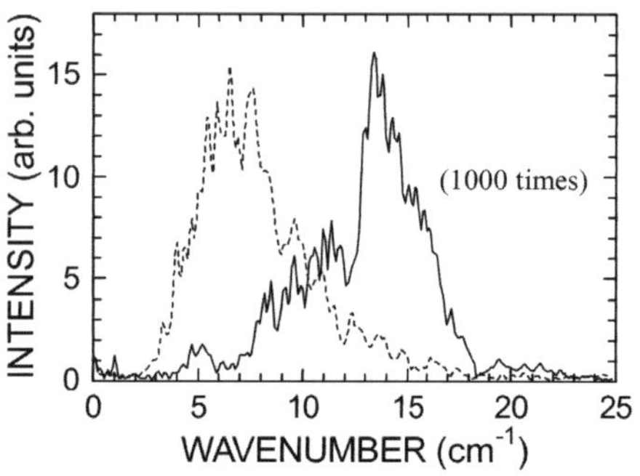

Figure 4. Observed spectra of CTR with (solid) and without (dashed) the illumination probe.

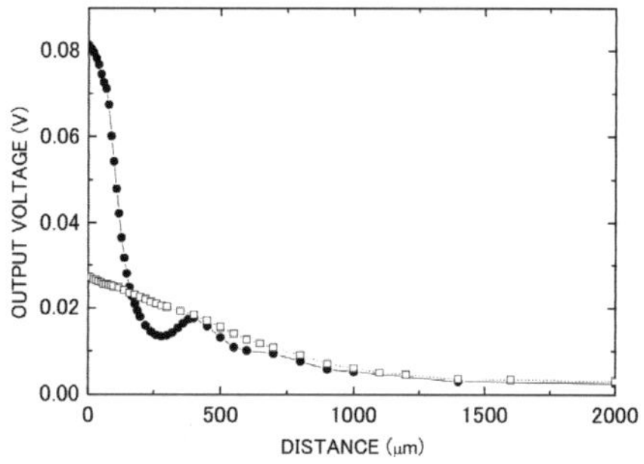

Figure 5. Dependence of observed intensity with the distance between the illumination and the collection probe. The solid circles and open squares show the results with the Kapton film and without it, respectively.

One of the theoretical models of the near field is the dipole-dipole interaction [8]. Therefore, the probe made of material with large refractive index has high efficiency of scattering the near field. In order to confirm the effect of the dielectric medium on the probe, the dependence of intensity with the distance between the illumination and the collection probe. The openings of the illumination and the collection probes are 2 mm and 200 μm in diameter, respectively. The detector equipped with the collection probe was moved to vary the distance. The solid circles connected by the solid line in Fig. 5 represent the experimental result using a sheet of Kapton film (polyimide film, Du Pont-Toray Co. Ltd) 25 μm thick on the top of the collection probe. The refractive index of Kapton and air are 1.7 and 1.0003 (sodium D line), respectively. The intensity is drastically enhanced within the distance equivalent to the size of aperture 200 μm. One of the possible reason of the dip around the distance of 250 μm is the Fraunhofer diffraction. If so, the dip should be also observed on the experimental result without the Kapton film. The reason of the dip is not clear at present. On the other hand, in the experimental result plotted by the open squares the Kapton film was not used, i.e. the collection probe had an empty opening only. The enhancement of intensity was not observed. The dielectric material with large refractive index on the probe has experimentally proved to scatter and detect the near field effectively.

Journal of Physics: Conference Series **359** (2012) 012016 doi:10.1088/1742-6596/359/1/012016

As the CTR has wideband spectrum, it is important to investigate the dependence of the spatial resolution power with wavelength experimentally. We have observed the dependence in the transmission mode. The conical cone of Teflon coated by silver paste on the tapered surface was used as an illumination probe. The size of the aperture was 450 μm. The collection probe with an opening 3mm in diameter was put on the detector. In order to measure the spatial resolution, the edge of a stainless steel plate 0.3 mm thick was scanned as a sample in front of the aperture on the top of the illumination probe. First, the spectra were measured at each scanning position of the sample. Second, the observed data were reconstructed to the scanning curves at each wavelength. The spatial resolution was estimated by the same way as Fig.3, i.e. the width of the first derivative of the scanning curve. The calculated spatial resolutions and resolution powers were summarized in Table 1. It was made clear that the resolution power became large with increasing the wavelength.

Table 1. Dependence of spatial resolution power with wavelength.

Wavenumber ν (cm^{-1})	Wavelength λ(μm)	Spatial Resolution Δx (μm)	Spatial resolution power $\lambda/\Delta x$
3.70	2700	371	7.3
4.94	2025	360	5.6
6.17	1620	341	4.8
7.41	1350	292	4.6
8.64	1157	243	4.8
9.88	1012	226	4.5
11.1	900.1	214	4.2
12.4	809.7	204	4.0
13.6	736.4	195	3.8
14.8	675.2	195	3.5
16.1	623.1	192	3.2
17.3	578.7	183	3.2

3.2. Reflection mode

Reflection measurement is also important for the THz-wave spectroscopy. The metallic conical cone with an aperture of 775 μm in diameter was used as an illumination probe. Two types of optical configuration were prepared to detect the reflected radiation. One was the beam-splitter type as shown in Fig.6 where the reflected radiation was divided by the Mylar beam splitter, focused by a Teflon lens, and guided to the detector. The other was the light-guide type as shown in Fig.7 where the reflected radiation was guided to the detector through the light pipe made of copper with an inside diameter of 10 mm.

In order to investigate the spatial resolution, the edge of a SUS plate 0.3 mm thick was scanned as a sample in front of the aperture on the top of the illumination probe. In the beam-splitter type configuration the observed intensity of the reflected radiation is plotted in Fig. 8 and the spatial resolution of 396 μm was derived from the Gaussian fitting of the first derivative curve. In Figure 9, the observed CTR spectrum with the sample is shown by open circles. The observed intensity remained even without the sample as shown by the dashed curve, which was considered to be the stray light reflected by the inside of the probe. The calculated net spectrum reflected by only the sample is represented by the solid curve. Considering the wavenumber of 5.7 cm^{-1} at the peak intensity of this spectrum, the spatial resolution power was evaluated to be 4.4.

Similarly, the spatial resolution and the spectra were measured in the light-guide type configuration. The scanning curve of the edge of a SUS plate is represented in Fig. 10. The intensity overshot around the edge. This phenomenon was not observed in the beam-splitter type configuration. The enhancement of intensity around the position of the edge is considered to arise from the stray light reflected at the cross-section face or the corner. In the beam-splitter type configuration the stray light is expected to stray off the optical path to the detector. On the other hand, it can be guided through the light pipe to the detector without a significant loss. The spatial resolution of 324 μm was obtained by the first derivative curve of Fig. 10. The spectra were shown in Fig. 11. The net spectrum subtracted the stray light without the sample (dashed curve) from the observed spectrum with it (open circles) is shown by the solid curve. Considering the wavenumber of 6.7 cm^{-1} at the peak intensity of this spectrum, the spatial resolution power was evaluated to be 4.6. This value is equivalent to that of the beam-splitter type.

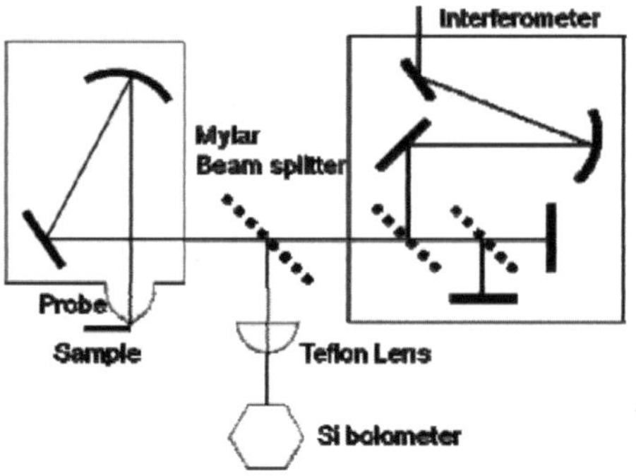

Figure 6. Schematic layout of the optics of the beam-splitter type.

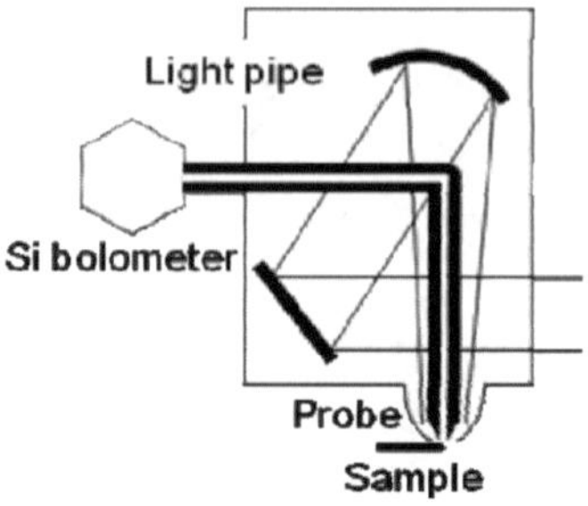

Figure 7. Schematic layout of the light-guide type.

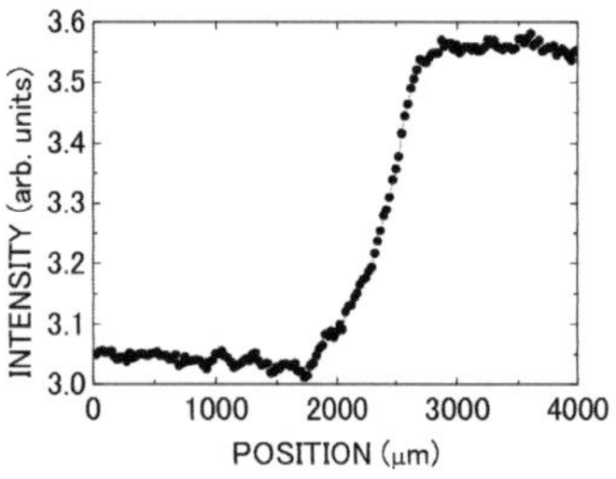

Figure 8. Scanning curve of the edge of a SUS plate 0.3 mm thick in the beam-splitter type configuration.

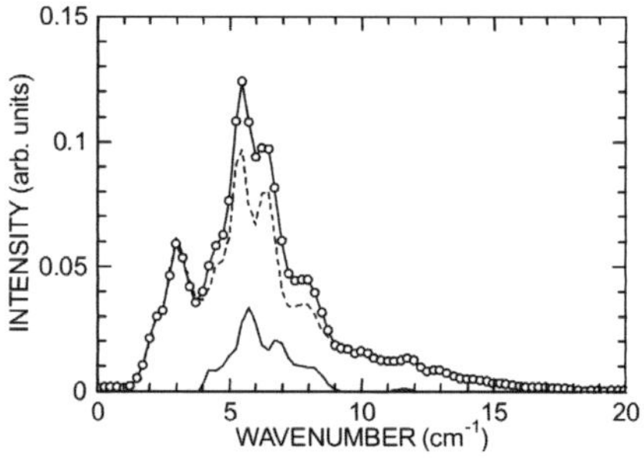

Figure 9. Observed spectra of CTR with (circles) and without (dashed) a sample. The solid curve shows the difference between the two spectra.

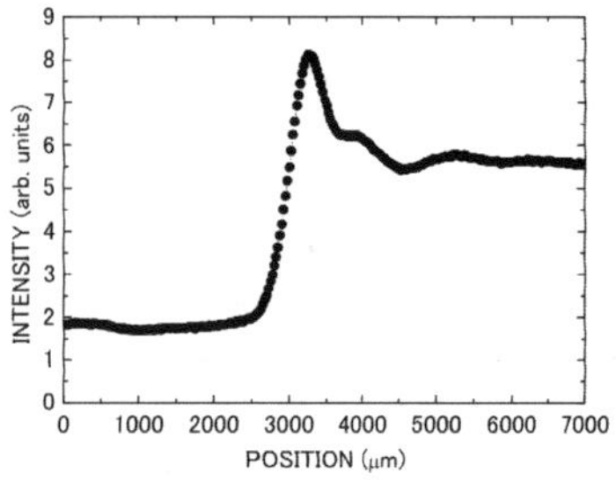

Figure 10. Scanning curve of the edge of a SUS plate 0.3 mm thick in the light-guide type configuration.

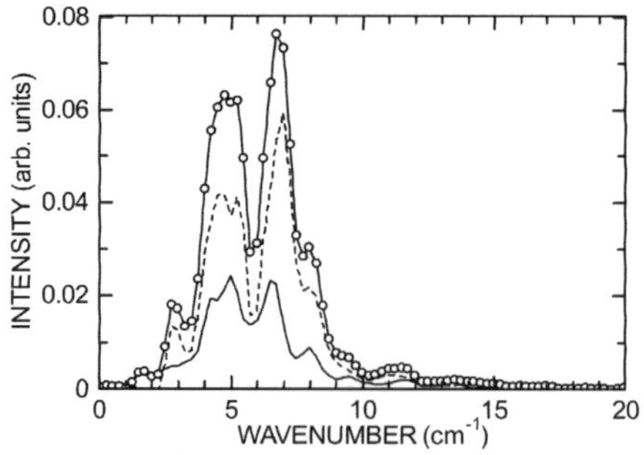

Figure 11. Observed spectra of CTR with (circles) and without (dashed) a sample. The solid curve shows the difference between the two spectra.

4. Summary

Average spatial resolution powers in accepting whole wavelengths of CTR were around 4.5 both in the transmission and reflection mode. The near field was effectively scattered and detected by usage of Kapton film with larger refractive index than air. The resolution power became large with increasing the wavelength. The scanning near-field THz-wave microscopy will be planned to be performed in the new THz-CSR beamline constructing at UVSOR. The average power of detected CTR without the aperture in the KURRI-LINAC was 0.5 μW. The pulse energy in a macro pulse and a micro bunch were 10 nJ and 160 pJ, respectively. The bandwidth of spectrum was wide around the frequency of 0.2 THz. At UVSOR the CSR is planned to be quasi-monochromatic with the pulse energy of 120 nJ at 1THz. Therefore the near-field imaging at the target frequency of absorption will be possible in the THz region.

Acknowledgement
This work was partly supported by Quantum Beam Technology Program of MEXT, Japan.

References

[1] Nakazato T, Oyamada M, Niimura N, Urasawa S, Konno O, Kagaya A, Kato R, Kamiyama T, Torizuka Y, Nanba T, Kondo Y, Shibata Y, Ishi K, Ohsaka T, and Ikezawa M 1989 Observation of Coherent Synchrotron Radiation *Phys. Rev. Lett.* **63** 1245.

[2] Shibata Y, Ishi K, Takahashi T, Kanai T, Ikezawa M, Takami K, Matsuyama T, Kobayashi K, and Fujita Y 1992 Observation of Coherent Transition Radiation at Millimeter and Submillimeter Wavelengths *Phys. Rev. A* **45** R8340

[3] Takahashi T, Kanai T, Shibata Y, Ishi K, Ikezawa M, Nakazato T, Oyamada M, Urasawa S, Yamakawa T, Takami K, Matsuyama T, Kobayashi K, and Fujita Y Cherenkov radiation from a finite trajectory of electrons 1994 *Phys. Rev. E* **50** 4041

[4] Shibata Y, Hasebe S, Ishi K, Takahashi T, Ohsaka T, Ikezawa M, Nakazato T, Oyamada M, Urasawa S, Yamakawa T, and Kondo Y 1995 Observation of coherent diffraction radiation from bunched electrons passing through a circular aperturein the millimeter- and submillimeter-wavelength regions *Phys. Rev. E* **52** 6787

[5] Shibata Y, Hasebe S, Ishi K, Ono S, Ikezawa M, Nakazato T, Oyamada M, Urasawa S, Takahashi T, Matsuyama T, Kobayashi K, and Fujita Y Coherent Smith-Purcell radiation in the millimeter-wave region from a short-bunch beam of relativistic electrons 1998 *Phy. Rev. E* **57** 1061

[6] Shade U, Holldack K, Kuske P, and Wustefeld G 2004 THz near-field imaging employing synchrotron radiation *Appl. Phys. Lett.* **84** 1422.

[7] Takahashi T, Matsuyama T, Kobayashi K, Fujita Y, Shibata Y, Ishi K, and Ikezawa M 1998 Utilization of coherent transition radiation from a linear accelerator as a source of millimeter-wave spectroscopy *Rev. Sci. Instrum.* **69** 3770

[8] Ohtsu M 1998 *Near-Field Nano/Atom Optics and Technology* (Springer-Verlag, Tokyo/Berlin) chapter 2

Fabrication and Characterization of Quasi-Optical Terahertz Nanorectifiers with Integrated Antennas

M. Ortolani[1,2], A. Di Gaspare[1], R. Casini[1,3], E. Giovine[1], S. Lupi[2] and V. Foglietti[1]

[1]*CNR-IFN – Institute for Photonics and Nanotechnologies, Rome, 00156 Italy*

[2]*Sapienza Università di Roma, Rome, 00185 Italy*

[3]*Università di Roma 2 Tor Vergata, Rome, 00173 Italy*

Michele.ortolani@ifn.cnr.it

Abstract. We are developing microelectronic terahertz rectifiers by scaling down the dimensions of GaAs Schottky diodes and field-effect transistors to the sub-micron range, and by investigating the effect of on-chip parasitic capacitances on the square-law power detection signal. For broadband operation at THz frequencies the terahertz oscillating signal is fed to the device by integrated lithographic planar antennas, suitably coupled to a silicon substrate lens. Such room-temperature THz detectors can be fabricated in arrays and naturally provide picosecond response time, suitable for detection of coherent THz radiation produced by single electron bunches in accelerators.

1. Introduction

Microelectronic rectifiers based on Schottky diodes and Field-effect transistors (FET) fabricated on epitaxial layers of III-V semiconductors like GaAs are commonly employed as square-law power detectors in millimeter wave integrated circuits (MMICs) working up to 100 GHz [1]. While terahertz Schottky diodes have been developed for mixing applications, it was recently demonstrated that FETs can be employed as terahertz detectors as well [2]. The working principle is based on the assumption that the non-linearity of current-voltage (I-V) characteristics exploited for diode and transistor operation at dc-to-gigahertz frequencies is somehow preserved in the sub-millimeter waves, where the concepts of power coupling and transit time delay are more properly used instead of simple I-V curves. In classical models [3] the square-law responsivity is directly proportional to the second derivative of the I-V curve, and inversely proportional to the dynamic resistance of the device (the inverse first derivative of the I-V curve). For these reasons, a good rectifier should display both highly nonlinear characteristics and low resistance at the bias point.

In this work, we demonstrate that by scaling the device size down to the nanometric range and by decreasing and properly controlling the on-chip parasitic capacitances, it is possible to preserve the rectifying behaviour of MMIC technology workhorses such as GaAs Schottky diodes and InGaAs/AlGaAs field-effect transistors up to 1 THz at least. For broadband detector operation at THz frequencies, the oscillating signal cannot be easily fed to the nanodevice by usual integrated microstrip or coplanar waveguides. Here, we fabricated the devices together with integrated lithographic planar antennas. The planar antenna can be suitably coupled to a silicon substrate lens pressed on the

Journal of Physics: Conference Series **359** (2012) 012017 doi:10.1088/1742-6596/359/1/012017

backside of the GaAs chip, and works as an optical-to-electronic converter: a free-space terahertz radiation beam, like that emitted by electron bunches in a synchrotron radiation beamline, is focused on the antenna by the silicon lens and the oscillating electric field is converted into a terahertz oscillating current, flowing trough the active device [4,5]. The responsivity of terahertz microelectronic detectors, also called *quasi-optical rectifiers*, mostly depend on the efficiency of such optical-to-electronic conversion, which is related to impedance matching issues, parasitic and intrinsic LC resonances and Gaussian beam quality [4].

We design the diodes and transistors with integrated antennas for terahertz operation and we fabricated them in our clean room with lithographic processes similar to those normally used by the microwave industry. The device quality and the process yield and reproducibility are not optimized, as our aim is rather the experimental investigation of the optical-to-electronic conversion. Selected devices with best dc characteristics are mounted in home-made packages with silicon lenses and we measured their responsivity in the 0.2-0.7 THz range with a tunable sub-millimeter electronic oscillator coupled to free space by a horn antenna.

At present, typical responsivities of quasi-optical rectifiers lie in the range 1-100 V/W, with a noise equivalent power around 1 nW/√Hz which can be favorably compared to that of classical room-temperature terahertz detectors such as pyroelectrics and Golay cells. As an advantage, quasi-optical rectifiers could be fabricated into arrays for beam-scan and beam-profiling, and naturally provide response time in the picosecond range (actually, as fast as the readout circuit allows), which is ideal for the detection of short THz radiation pulses produced by single electron bunches in accelerators [5].

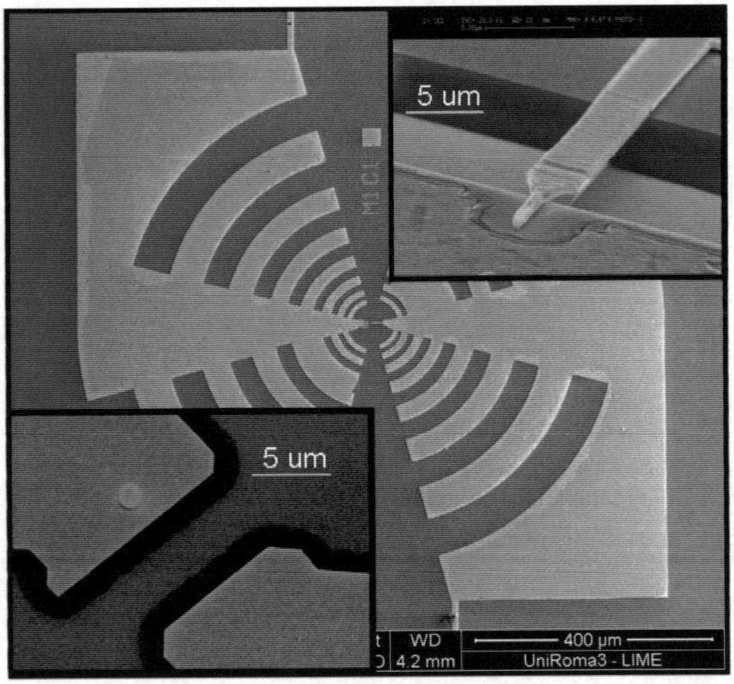

Figure 1. Scanning Electron Microscope image of an integrated log-periodic antenna fabricated by electron-beam lithography in the CNR-IFN clean room. The nominal antenna bandwidth is 0.1-1.0 THz, as the teeth radius ranges from 350 to 35 microns. At the antenna feed, either an air-bridge Au/Ti/n-GaAs Schottky diode (upper right inset) or a high-electron-mobility transistor with pseudomorphic InGaAs channel (lower left inset) are fabricated. In the case of the transistor, the gate electrode is added in a subsequent step (not shown).

Journal of Physics: Conference Series **359** (2012) 012017 doi:10.1088/1742-6596/359/1/012017

2. Design and Fabrication of Schottky Diode Rectifiers

The Schottky diode based on epitaxial electron-doped gallium arsenide (n-GaAs) is probably the most important microelectronic device for operation in the sub-millimeter and terahertz range [1]. Diode areas of the order of 1 square micron are needed to obtain a low junction capacitance $C_j \sim 10^{-15}$ F, which was classically achieved by pressing etched tungsten whiskers on GaAs chips [3]. In the last decade, air-bridge lithographic anodes have been developed (see Fig. 2A), allowing for multiple anodes on–chip, higher yield of the fabrication process, and studies of the physical properties of the Schottky junction at terahertz frequencies. The prominent role played by epitaxial n-GaAs diodes in sophisticated terahertz spectroscopic instruments such as heterodyne mixers [1,4,5] and frequency multiplied oscillators is due to the simultaneously high electron mobility and density (up to 3000 cm^2/Vs and $5*10^{18}$ cm^{-3} at room temperature, respectively) obtainable in GaAs, which ultimately provide a low series resistance R_s and hence a high cutoff frequency $f_c = 1/2\pi R_s C_j$. More recently, the development of terahertz imaging and telecommunications points towards the development of fast terahertz power detectors, producing a rectified signal modulated up to the nanosecond range [5].

We are developing Schottky diodes with high doping level of the semiconductor ($N_d > 10^{18}$ cm^{-3}) for operation at THz frequencies. The high doping is crucial for very high frequency operation, since it provides a small series resistance and a short dielectric relaxation time [6]. However, in highly doped metal-semiconductor junctions the tunneling current through the quasi-triangular potential barrier contributes substantially to the total current, even at room temperature [7]. This is shown in Fig. 2 where we performed the calculation of the potential barrier seen by electrons in the semiconductor (Fig. 2a) and of the tunneling coefficient T(ε) as a function of the distance in energy ε from the top of the barrier by using the triangular-barrier approximation (Fig. 2b). Temperature was set to 295 K and standard Fermi distribution functions were employed for the metal (Au) and the semiconductor (GaAs). The doping level of GaAs was varied from $N_d = 10^{17}$ cm^{-3} which is normally used for microwave rectifiers and displays negligible tunneling contribution (dashed lines in Fig. 2) to $N_d = 3 \cdot 10^{18}$ cm^{-3} which is at the verge of the transition to a completely symmetric low resistance junction (i.e. an Ohmic contact to the semiconductor, thick lines in Fig 2) and hence cannot be used as rectifier. Our choice of the doping level for the GaAs epitaxial material was therefore $N_d = 1 \cdot 10^{18}$ cm^{-3}.

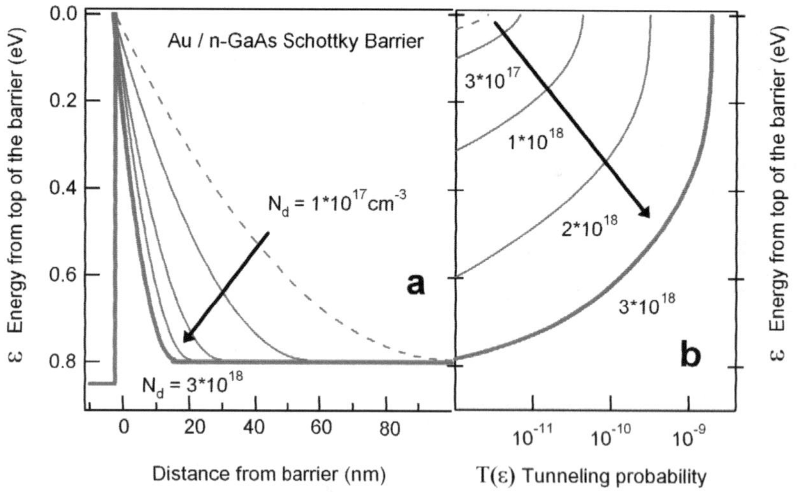

Figure 2. (a) Potential barrier seen by electrons in GaAs as a function of the distance from the Au/GaAs Schottky junction. (b) Differential tunnelling coefficient as a function of energy distance from the top of the barrier (abscissa and ordinate axis are reversed on purpose). The triangular-barrier approximation is used for the calculation in (b).

Journal of Physics: Conference Series **359** (2012) 012017 doi:10.1088/1742-6596/359/1/012017

The typical signature of tunneling contribution in the I-V characteristics shown in Fig. 3 for our and other Schottky diodes is an ideality factor η different from unity. Indeed, in forward-bias the tunneling current I_{tun} has the same exponential dependence on the bias voltage V_b as the thermionic current $I_{th} = I_{sat} \cdot \exp(V_b / kT)$, because tunneling essentially takes place close to the top of the barrier, where the value of the Fermi distribution in the semiconductor $F_s << 1$ can be approximated by $F_s = \exp[(V_b - \phi_B)/kT]$, and $I_{tun} = T_\varepsilon \cdot \exp(-\phi_B / kT) \exp(V_b / kT)$ where T_ε is the tunneling probability integrated from the top of the barrier to a given value ε and ϕ_B is the barrier height (0.85 V in Au/n-GaAs). The total current can then be expressed in the usual compact form

$$I=I_{th}+I_{tun}= I'_{sat}*[\exp(V_b/\eta kT) -1] \qquad (1)$$

which defines the ideality factor η. Therefore, the main effect of the tunneling current is a modification of the exponential characteristic voltage, conventionally expressed in terms of $\eta > 1$. The I-V plot of our standard diodes in the forward-bias region (Fig. 2b, $V_b >$ 0.4) indeed shows an exponential dependence with $\eta = 1.7$ and a saturation current much larger than that of diodes of equal area fabricated on GaAs epilayers with $N_d = 10^{17}$ cm^{-3} [5], hence confirming that tunneling plays a key role in our devices and that classical Schottky theory (Eq. 1) describes well the I-V curve of our standard diodes.

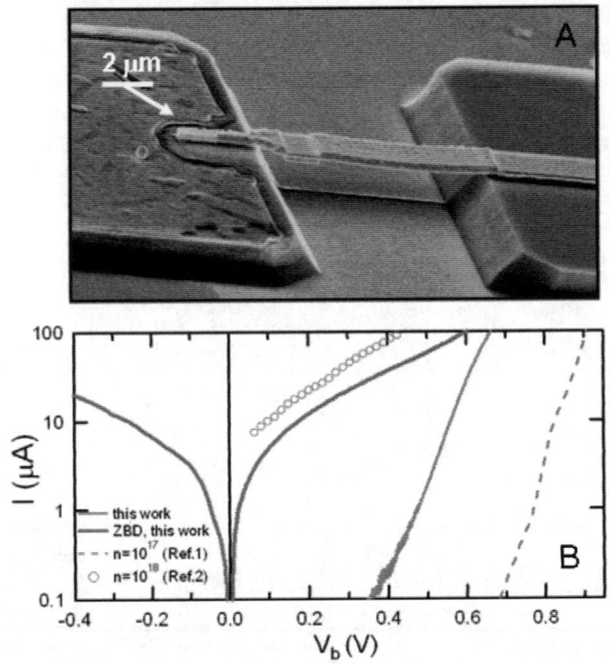

Figure 3. (A) SEM micrograph of a 0.2x2 µm^2 air-bridge Schottky diode fabricated at CNR-IFN in Rome, Italy. The 2-micron long extremity of the bridge indicated by the arrow is the T-section Au anode contacting the highly doped GaAs surface. (B) I-V curves of Schottky diodes with an area of 2 µm^2. Red: standard diode, this work ($N_d = 10^{18}$ cm^{-3}); blue: Zero-Bias Diode, this work; dashed green, circular mixer from Ref. 5 ($N_d = 10^{17}$ cm^{-3}); empty dots: T-section mixer from Ref. 1 ($N_d = 10^{18}$ cm^{-3})

Let us now focus on the I-V curves of Fig. 3b close to zero bias. I-V measurements were performed with a dc-current bias, which is the configuration used to bias rectifying detectors. In the case of the standard diode, the minimum bias current I=100 nA corresponds to V_b =0.37 V, as no such current can flow in the diode without a significant applied voltage. Recently, we have also successfully fabricated Zero Bias Diodes (ZBD) with a very similar process where we have additionally shaped the three-

Journal of Physics: Conference Series **359** (2012) 012017 doi:10.1088/1742-6596/359/1/012017

dimensional junction geometry to further decrease the Schottky barrier width [9]. In these diodes, the minimum bias current flows for $V_b < 10$ mV, displaying quasi-ohmic I-V relation around the zero-bias point, not described by Eq. 1. From now on, we shall define the dynamic junction resistance, mainly due to tunneling, at the bias point

$$R_d = (dI/dV)^{-1} \; @ \; V_b \qquad (2)$$

by keeping in mind that the I-V curves are still non-linear at any V_b, including $V_b = 0$: this is of course crucial for the intended use of the diode as THz radiation rectifier. Together with the R_s and C_j, R_d defines the zero-bias cutoff frequency of the rectifier f_{ci} , which is derived by calculating the transfer of power from the diode to an external load equal to R_d [3]:

$$f_{ci} = 1/2\pi \; C_j(R_sR_d)^{0.5} \qquad (3)$$

If we take $R_s \sim 20$ Ohm (a typical value for highly doped epitaxial n-GaAs diodes) and $C_j = 1$ fF, zero-bias rectification at THz frequencies requires R_d as low as possible. As seen from Fig. 3, for standard diodes with sub-micron area, $R_d > 10^9$ Ohm at zero bias, therefore the diode needs to be forward-biased to work as THz detector. Instead, for the ZBDs we obtain minimum values for R_d around 100 kOhm for 0.4 um^2 diodes at $V_b = 0$ which indeed allow for THz detector operation without any applied dc bias [9], a clear advantage for future multi-pixel arrays.

In summary, tunnelling currents through a Schottky junction realized on a highly doped GaAs layer provide low dynamic resistance as shown in Fig. 4a, by still keeping a nonlinear dependence of the tunnelling coefficient as shown in Fig. 4b for a realistic value of the peak-to-peak THz voltage applied to the diode through a planar antenna in quasi-optical mode.

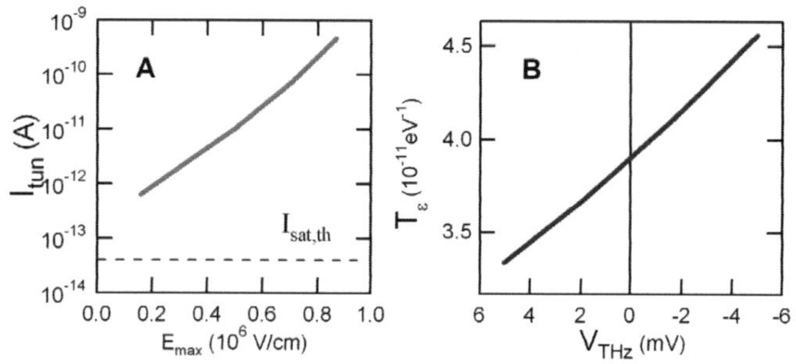

Figure 4. (A) Tunneling current (Eq. 1) in a 0.2x2 μm^2 Schottky junction calculated in the triangular-barrier approximation, as a function of the constant electric field in the depletion region. Note the exponential dependence on the electric field. (B) Tunneling probability T_ε integrated over the Schottky barrier for a Au/GaAs junction with high doping level of $N_d = 10^{18}$ cm^{-3} as a function of a small applied voltage V_{THz}, of magnitude comparable to that produced by realistic terahertz beam (c.w. power of 1 mW and diffraction-limited focussing are assumed). Note the dependence of the tunnelling probability on V_{THz} which ensures partial rectification at zero dc bias.

We designed the n/n+ epitaxial structure according to calculations shown in Fig. 2 (100 nm with $N_d = 1\cdot10^{18}$ cm^{-3} on top of 1000 nm with $N_d = 5\cdot10^{18}$ cm^{-3} for ohmic contact to the cathode). Layers were grown by Molecular Beam Epitaxy on a semi-insulating GaAs wafer. On these wafers, we have fabricated antenna-coupled terahertz Schoottky diodes in our clean room. As shown in the scanning-electron micrograph of Fig. 3a, the sub-micrometric Schottky junction is achieved by contacting the GaAs surface with the extremity of a free-standing metal bridge, with mesa isolation between anode and cathode pads to cut parasitic capacitances. The layout of the alloyed ohmic contacts was identical in both cases, and it was designed to work as an integrated planar dipole antenna, centered at 0.22

Journal of Physics: Conference Series **359** (2012) 012017 doi:10.1088/1742-6596/359/1/012017

THz. The sub-micron metal-semiconductor junction was realized with the help of an electron-beam writer, by patterning rectangles at different doses on a trilayer resist polymer based on polymethylmethacrilate (PMMA), so to obtain a "T-section" anode [9] with lateral footprint of only 0.2 microns, while the anode length (2 to 10 μm) was designed to obtain diodes of different area (0.4 to 2 μm^2). After patterning, in standard diodes surface oxides are first removed by immersion of the wafer in a NH_4OH solution and then the metal contact is evaporated (30 nm Ti and 400 nm Au). After the mesa-etching step, which also suspends the air-bridge, wafers are dc-tested before dicing.

3. Terahertz Rectification Measurements

Terahertz rectification measurements took place at University of Rome "Sapienza" by using the radiation source therein installed in the framework of the SPARC linear accelerator project. The devices are diced in 3x3 mm chips, containing many diodes, whose polished backside is put in contact with the flat side of a hyper-hemispherical Silicon lens, which focuses an incoming radiation beam onto the diode. The contacts work as a $\lambda/2$ dipolar antenna resonating at 0.25 THz. The distance between two neighboring devices is only 320 μm, which allows the chip to be mounted in the focus of a single Si substrate lens (the effective wavelength λ_{eff} = 450 μm defines the focal spot size in case of ideal diffraction-limited focussing). For responsivity measurements at 0.27 THz we have used the quasi-optical setup shown in Fig. 5, based on a Gunn oscillator followed by a frequency multiplier (270GHz-Tx by Virigina Diodes Inc.) and by a horn antenna. The continuous-wave beam at 0.27 THz is mechanically chopped at 330 Hz and sent to the silicon lens. The detector box contains an analog video amplifier with bandwidth of 1 kHz and gain of 1000, ac-coupled and directly bonded to the diode chip, together with a bias-tee which allows dc-bias by batteries, adjustable with a potentiometer . In this way, the pick-up of ambient noise can be neglected and the signal can be safely input in a standard lock-in amplifier.

Figure 5. Compact terahertz generator installed at University of Rome "Sapienza" used for rectification measurements on Schottky diodes and calibration of other detectors. Radiation at 0.27 THz is emitted by a Gunn diode at 90 GHz followed by a frequency tripler (by Virginia Diodes Inc.). A horn antenna is used for free-space coupling into a Gaussian beam, which is then directed to a mechanical chopper and to the detector to be tested (right part of the picture).

The diode responsivity is calculated by dividing the lock-in signal by the video amplifier gain and by the radiation power reaching the diode, estimated around 0.01 mW by calculating the solid angle overlap between the horn antenna and the silicon lens patterns. The results are shown in Fig. 6 and display a very good agreement with the theoretical responsivity calculated from the measured non-linearity of the dc characteristics. The responsivity was found to be independent from the chopper frequency in the 10 Hz – 550 Hz range.

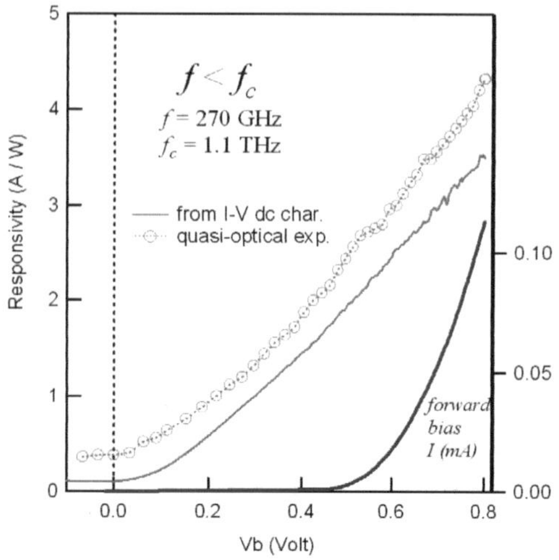

Figure 6. Terahertz responsivity of a GaAs Schottky Diode with junction area of 2 square microns, produced at CNR-IFN and measured at University of Rome "Sapienza" as a function of V_b. Red circles: rectified signal when the packaged diode is exposed to 0.27 THz radiation. Note that the diode cutoff frequency according to Eq. 3 is 1.1 THz. Green line: response calculated from the I-V curve (blue line, right axis) in Fig. 3 following the model of Ref. 3. The agreement is very good, apart from the nonzero response signal found at zero bias, which is due to asymmetric tunnelling (see Fig. 4 and related discussion).

4. Heterostructure Field Effect Transistor Rectifiers

InGaAs-channel HFETs with gate length of 1 micron and channel width of 10 microns were fabricated with integrated bow-tie antennas (Fig.1). For the device described in this paper, the antenna couples the radiation polarized parallel to the channel direction to the carriers in the two-dimensional electron gas (2DEG). Out-of-phase self-mixing with the gate voltage (possibly induced by capacitive coupling among contact pads and/or the gate electrode itself) produces the rectification of the source-drain current inside the channel, and a down-converted output can be measured at low-frequency at the drain port. Details of the mixing mechanism, including the role of collective (plasma) oscillations have been discussed in several works [2, 10].

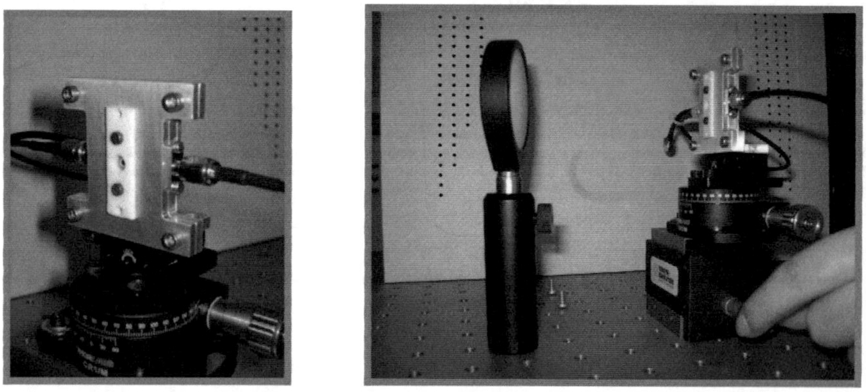

Figure 8. Heterostructure Field Effect Transistor Detector mounted in a Quasi-optical package for terahertz rectification measurements developed at CNR-IFN. The Detector has two SMA bias/readout ports (gate and drain contacts) and the radiation is coupled through a hyper-hemispherical silicon lens (at the center of the white Teflon spring) and an on-chip planar lithographic antenna (see Fig. 1).

In the present device, we made an effort to optimize the HFET detector sensitivity by setting the maximum responsivity close to zero gate bias V_g. In Fig. 7 we show the transfer characteristics of the transistor (drain current at fixed drain voltage bias I_d vs.V_g) displaying a threshold voltage of V_{th} = -200 mV. This means that the 2DEG in the channel is partially depleted at $V_g = 0$ and therefore strongly non-linear I_d-V_g characteristics are expected also at THz frequencies with no dc bias applied at the gate. This simplifies the detector circuit, in a way similar to what happens in the ZBD Schottky detectors of section 2 [9].

To measure the value of the rectified signal, the HFET chip was packaged together with a silicon lens and it was exposed to the radiation generated by a frequency-multiplied tunable electronic oscillator in the range 0.15-0.75 THz (by Virginia Diodes Inc.), emitted in free-space by a horn antenna, collected and refocused onto the detector lens by a pair of 90° off-axis parabolic mirrors. The emitted power was independently measured by a calorimeter to account for variation of the emitted power with frequency. The oscillator output is amplitude-modulated at 1 kHz by a master oscillator, which also works as reference for a lock-in amplifier attached to the drain port to measure the response signal.

Full spectroscopic characterization will be reported elsewhere [11]. As an example, we show in Fig. 8 the dependence of the response signal of one of the fabricated HFET to radiation at 510 GHz as a function of V_g. The peak responsivity value is found at the pinch-off voltage, as described in Ref. 2 and as expected from the higher nonlinear coefficient in that voltage range. As seen from the red curve, the result of our design and fabrication efforts is that almost 60% of the maximum response signal is obtained at $V_g = 0$. A quantitative comparison of noise and responsivity of Schottky and HFET detectors is ongoing [10].

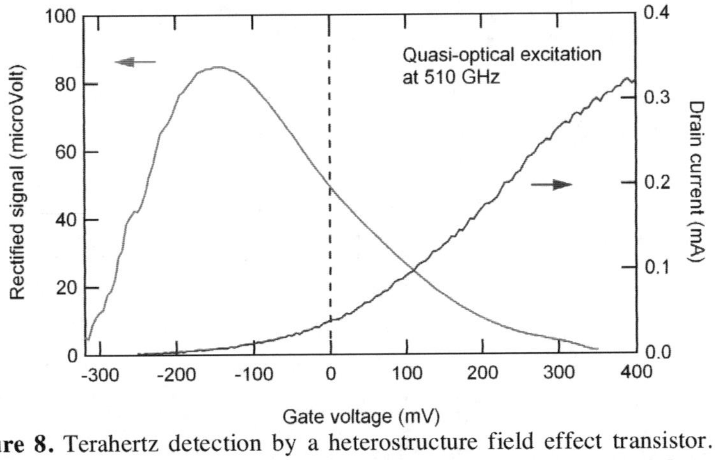

Figure 8. Terahertz detection by a heterostructure field effect transistor. Transfer characteristics (blue curve, right axis) and response signal to 510 GHz radiation (red curve, left axis) as a function of the dc gate bias voltage. The maximum response signal is found at pinch-off.

This work was partially supported by Italian Ministry of Research (MIUR) programs FIRB Italnanonet grant no.RBPR05JH2P_014 and FIRB "Futuro in Ricerca" grant no. RBFR08N9L9.

References

[1] Crowe T W, Porterfield D W, Hesler J L, Bishop W L, Kurtz D S and Hui K 2005 *Proc. SPIE* **5790** 271
[2] Knap W *et al* 2002 *J. Appl. Phys.* **91** 9346
[3] Cowley A M and Sorensen H O 1966 *IEEE Trans. Microw. Theory Tech.* **14** 588
[4] Hesler J L, Liu L, Xu H, Duan Y, Weikle R M 2008 *IRMMW-THz 2008, 33rd Intl. Conf. on Infrared, Millimeter and Terahertz Waves*

[5] Semenov A, Cojocari O, Hübers H W, Song F, Klushin A and Müller A S 2010 *IEEE Electron Device Letters* **31** 7
[6] Guetin P and Schreder G 1986 *J. Appl. Phys.* **42** 5689; Zussman A 1986 *J. Appl. Phys.* **59** 3894
[7] Van Der Ziel A 1977 *J. Appl. Phys.* **47** 2059
[8] Chahal P, Morris F, and Frazier G 2005 *IEEE Electron Device Lett.* **26** (12) 894
[9] Casini R, Di Gaspare A, Giovine E, Notargiacomo A, Ortolani M and Foglietti V 2011 submitted to *Appl. Phys. Lett.*
[10] Di Gaspare A, Casini R, Ortolani M, Giovine E, Foglietti V, Romanini P, Dominijanni D, Peroni M and Evangelisti F 2011 *IEEE Proc. of IRMMW-THz*
[11] Di Gaspare A, Casini R, Giovine E, Foglietti V, Evangelisti F and Ortolani M, manuscript in preparation.

Journal of Physics: Conference Series **359** (2012) 012018 doi:10.1088/1742-6596/359/1/012018

The THz Radiation Source at the SPARC Facility

E. Chiadroni, A. Bacci, M. Bellaveglia, M. Castellano, G. Di Pirro, M. Ferrario, G. Gatti, E. Pace, A. R. Rossi, C. Vaccarezza
INFN/LNF, via E. Fermi 40, 00044 Frascati, Italy

P. Calvani, A. Nucara, D. Nicoletti, O. Limaj, S. Lupi
Phys. Dept. and INFN University of Rome "La Sapienza", p.le A. Moro, 1 - Rome, Italy

B. Marchetti, A. Cianchi
Phys. Dept. and INFN University of Rome "Tor Vergata", via della Ricerca Scientifica 1, 00133 Rome, Italy

A. Mostacci
SBAI Dept. University of Rome "La Sapienza", Rome, Italy

C. Ronsivalle
ENEA C.R. Frascati, Italy

A. Perucchi
Elettra Synchrotron, SS 14, Basovizza, Trieste, Italy

E-mail: `enrica.chiadroni@lnf.infn.it`

Abstract. The interest for Terahertz (THz) radiation is rapidly growing, both as it is a powerful tool for investigating the behavior of matter at low energy, and as it allows for a number of possible spectroscopic applications spanning from medical science to security. The linac-driven THz source at the SPARC facility can deliver broadband THz pulses with femtosecond shaping and can be used for electron beam diagnostics to fully reconstruct the longitudinal charge distribution. Beyond this application, the possibility to store much more energy in a single THz pulse than table-top sources renders the SPARC THz source very interesting for a spectroscopic use. In addition, taking advantage from electron beam manipulation techniques, high power, narrow-band THz radiation can be also generated. Those source characteristics provide a unique chance to realize THz-pump/THz-probe spectroscopy, a technique practically unexplored up to now.

1. Introduction

SPARC is an R&D photoinjector facility for the generation of high brightness electron beams, which was originally conceived to perform electron beam dynamics studies and Free Electron Laser (FEL) experiments, both under Self Amplified Spontaneous Emission (SASE) regime in the

Journal of Physics: Conference Series **359** (2012) 012018 doi:10.1088/1742-6596/359/1/012018

visible range at 500 nm [1], [2] and down to few tens of nm wavelengths, in the non conventional schemes such as seeded FEL, cascade FEL, high harmonics generation in gas [3], [4]. Recently, the SPARC facility is extending its prospects to a wider range of applications, e.g. Plasma and Laser Wakefields Acceleration (PWFA-LWFA), Thomson scattering and high peak power THz radiation.

The motivation for the development of a THz source at SPARC lays in the ever growing interest in filling the so-called THz gap with a high peak power radiation, typical of a linac-based source. Indeed, from simulations the expected peak power for the SPARC THz source is foreseen in the order of 10^8 Watts (Fig.1, right plot). Such a power performance would produce pulses delivering an energy in the order of few (up to ten) μJ (Fig.1, left plot), well above THz radiation from synchrotrons and most standard table top sources. Applications of high

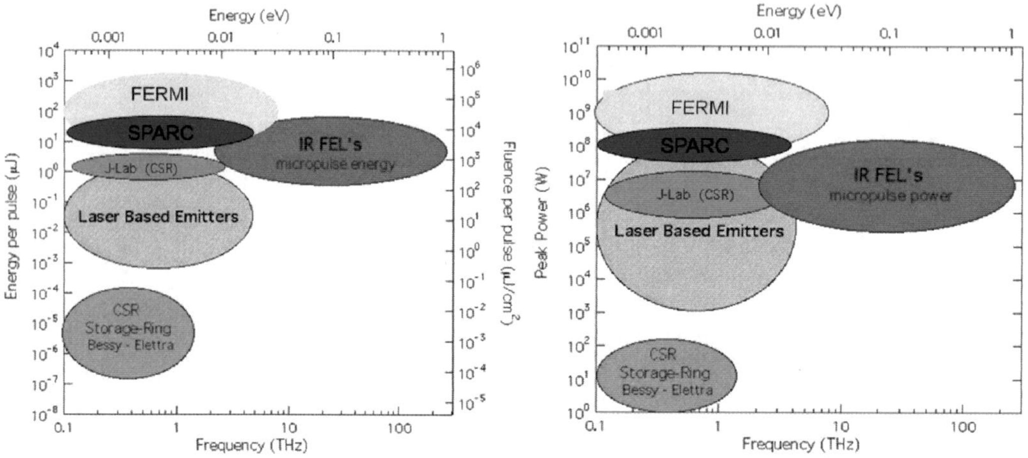

Figure 1. Energy/fluence per pulse (left) and peak power (right) figures of merit of the THz CTR emission at SPARC are plotted in comparison to those of conventional THz sources.

peak power pulses concern mainly time domain spectroscopy with THz pump/THz probe and frequency domain measurements on novel materials, e.g. superconductors.

From an accelerator point of view, coherent THz radiation is a valid tool for electron beam longitudinal diagnostics, enabling to reconstruct the longitudinal charge distribution of the bunch.

2. Theory

Radiation extending up to the THz range can be generated as coherent radiation from ultra-short ($\approx$ sub-ps) electron bunches. Any radiation process which does not change the electron distribution during the emission, and in which no interaction between particle and radiation happens, can be used to retrieve the longitudinal electron beam charge distribution by studying the coherent spectrum of radiation emitted by the bunch. This is true, in particular, for transition and diffraction radiation, where the emission is a surface phenomenon and thus virtually instantaneous; however, if the electron distribution in the bunch can be considered frozen during the emission, also synchrotron radiation, spontaneous undulator radiation and Cherenkov radiation may be considered as suitable sources.

Journal of Physics: Conference Series **359** (2012) 012018 doi:10.1088/1742-6596/359/1/012018

2.1. Coherent Radiation

The total radiation intensity emitted by a bunch of electrons, $I(\lambda) = I_{sp}[N + N(N-1)F(\lambda)]$, can be interpreted as the sum of two terms [5]: one, proportional to the number of particles N, represents incoherent emission, whereas the one proportional to N^2 implies a coherent process, which dominates the emission at wavelengths (λ) equal and longer than the bunch length. I_{sp} is the single particle radiation intensity, which depends on the emission mechanism considered. The coherent intensity depends on the bunch longitudinal dimension through the longitudinal form factor, $F(\lambda)$ [6], defined as the Fourier transform of the bunch longitudinal profile. It is then clear that, by measuring the coherent spectrum, the form factor is determined and the longitudinal profile can be retrieved.

2.2. Transition Radiation

Let us consider as emission phenomenon the radiation produced when a relativistic charged particle crosses the boundary between two media with different dielectric constants, *e.g* a vacuum-metal screen interface. The transition radiation (TR) originates from the time varying currents induced, on the target surface, by the field of the traveling charge, due to optical inhomogeneities in the space. Assuming a perfectly conducting, infinitely thin target, which extends indefinitely transversally, the spectral and angular energy distribution from a single electron is well described by the Ginzburg-Frank formula in far field approximation [7]

$$I_{sp} = \frac{e^2}{4\pi^3\varepsilon_0 c}\frac{\beta^2\sin^2(\theta)}{(1-\beta^2\cos^2(\theta))^2} \tag{1}$$

with $\beta = v/c$ and θ the emission angle. In such picture, the TR spectral distribution of radiation frequencies well below the target plasma frequency, ω_p, does not depend on frequency itself; moreover the intensity is concentrated within a cone whose axis is coincident with the particle velocity and displays an angular aperture proportional to $1/\gamma$ (γ being the Lorentz factor). In a real experimental setup, the assumptions of infinite target size and far-field approximation are not usually fulfilled, causing the TR spectrum to become a complicate function of beam energy, target extension, frequency and angle of emission [8]. Since the natural extent of the particle transverse field Fourier components is of the order of $\gamma\lambda$, when the electron energy is relatively high long wavelength components may exceed the dimension of the target, resulting in a strong suppression of low-frequencies. Indeed, as it is shown in Fig.2, for a 90 MeV particle energy and a screen radius of 20 mm, even at 1 THz the real spectral angular distribution is not well reproduced by the Ginzburg-Frank formula.

2.3. Beam Manipulation Techniques

Short electron bunches are mandatory for the generation of high peak power, broad band, extending up to high frequencies, THz radiation. In order to longitudinally compress the beam down to few hundreds of femtosecond, the velocity bunching (VB) [10] is used at SPARC [11]. The VB technique relies on the fact that the beam is injected in the first accelerating structure at the phase of zero crossing RF field. Since the beam is slightly slower than the phase velocity of the RF wave, it slips back to phases where the field is accelerating, and it is simultaneously chirped and compressed. Due to non-linearities in the beam longitudinal phase space, the longitudinal profile of the compressed beam has a sharp rising edge, which is responsible of emission at higher THz frequencies.

As an alternative to broadband THz radiation, coherent narrow-band, high peak power THz radiation can be produced by means of longitudinally modulated electron beams generated by tailoring the shape of the drive laser pulse. Several techniques can be considered, e.g the one presented in [12]; here we will describe the one in use at SPARC [13], [14].

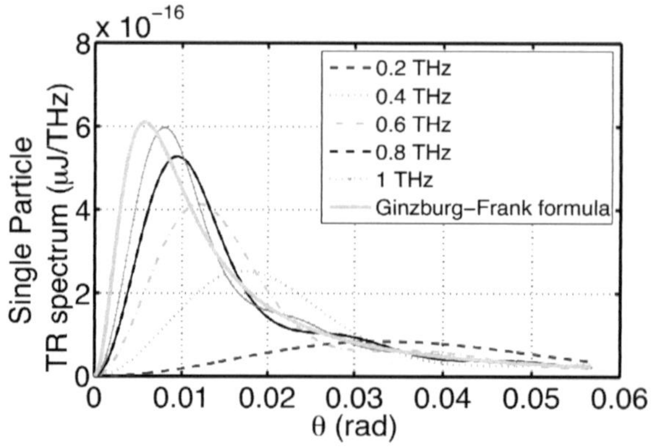

Figure 2. Modified Ginzburg-Frank formula [9] to take into account the effect of a finite size target (with radius a=20 mm) with respect to the natural size of the particle field ($\gamma\lambda$, γ=176). The spectral angular distribution for different wavelengths is compared to the ideal Ginzburg-Frank formula.

If a train of n laser pulses, named as laser comb, with picosecond distance, illuminates the photocathode, assuming an instantaneous emission, each pulse will produce a disk of electrons with the same spacing as the laser pulses. Due to longitudinal space charge forces, the bunch lengthens and the longitudinal modulation is mixed up at the end of the drift downstream of the gun. Nevertheless, memory of the original structure is kept in the longitudinal phase space, manifesting in a saw-tooth like energy profile. Thus, in the VB process, due to a correlated time-velocity chirp in the electron bunch, causing electrons on the tail to be faster than those on the head, a rotation in the longitudinal phase space occurs in the first accelerating section and the longitudinal modulation is restored at the end of the linac.

If a comb train with THz repetition rate is transported to the TR target, narrow band emission occurs at the frequency corresponding to the pulse inter-distance. The longitudinal form factor of a comb-like electron beam distribution, as that shown in Fig.3, is a single peak at the comb repetition rate, whose intensity is the same as if all the electrons were confined in the single sub-pulse. In case of shorter electron pulses, higher and more intense harmonics show up in the comb form factor, as pointed out in Fig.4.

The emission frequency can be tuned by changing the distance between pulses in the train. Therefore combining the VB technique and a comb-like electron beam distribution, a tunable narrow band THz source can be produced.

3. Experimental Layout

Coherent Transition Radiation (CTR) is generated at the interface of an Aluminum coated Silicon screen. The screen is placed at the end of the by-pass line (Fig.5) in the vacuum pipe at 45° with respect to the electron beam direction. CTR is then extracted at 90° through a z-cut quartz window. Radiation is then collected by a 90° off-axis parabolic mirror. Parallel radiation is then reflected down to a flat mirror at 45° which reflects radiation horizontally. Two schemes are foreseen (Fig.6): one for interferometer measurements and one for integrated CTR measurements with the possibility of selecting custom band pass filters in the THz range [15].

A Martin-Puplett interferometer [16] is used for frequency domain measurements. The main advantages of this device derive from the beam splitter properties: it splits polarizations and, being made of wires, reflectivity coefficients depend on wires diameter and spacing. Furthermore,

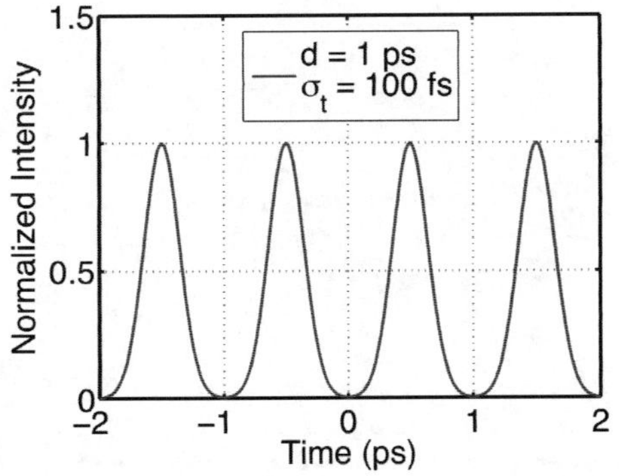

Figure 3. Calculated four-pulses comb profile with 1 ps pulse separation and 100 fs rms sub-pulse length.

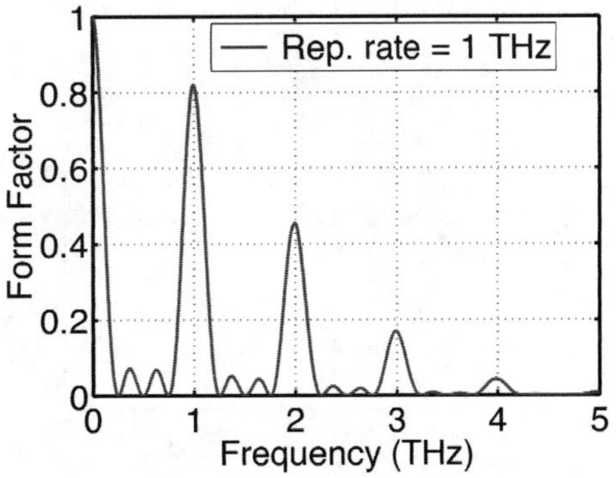

Figure 4. Form factor corresponding to the four-pulses comb train: narrow band THz emission occurs at the comb repetition rate, i.e. 1 THz, and at higher harmonics.

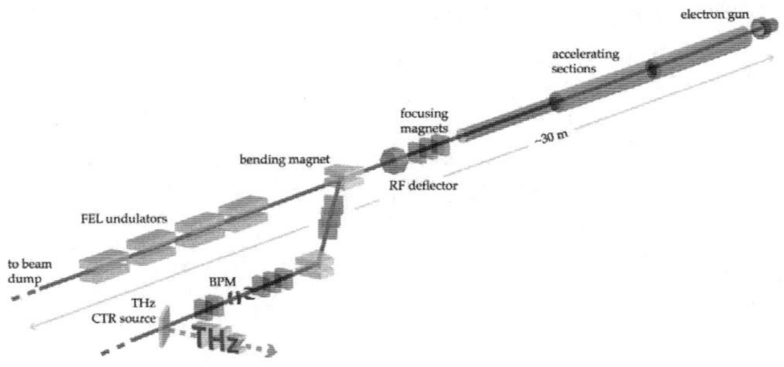

Figure 5. Layout of the SPARC accelerator, with the THz source placed at the end of the dogleg line.

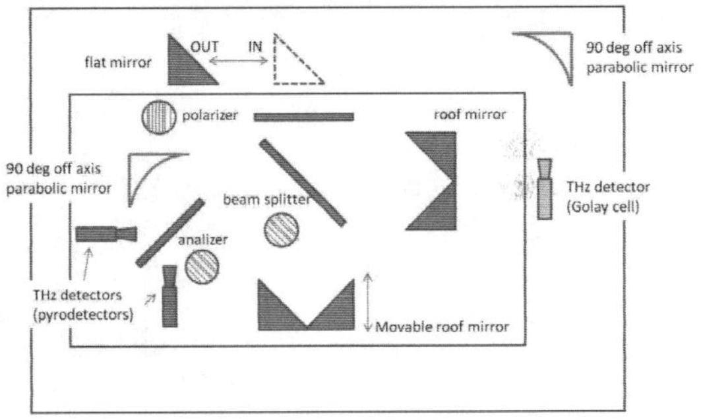

Figure 6. Experimental layout for detection of THz radiation (top view).

measuring the intensity of both polarizations, the difference interferogram can be normalized to the sum, thus canceling correlated fluctuations due to electron beam instabilities.

4. Results

High peak power ($\approx$ MW), broad band THz radiation up to 5 THz has been measured at SPARC with a 500 pC, 500 fs electron beam. Details can be found in [17].

Narrow band THz radiation has been produced and characterized by studying the rotation of the longitudinal phase space of a four-pulses train under VB regime. At the maximum compression phase, the four pulses completely overlap in time (Fig.7); moving the compression phase toward the over compression regime, the bunches get better separated and modulated (Fig.8).

The relevant regime for THz applications, i.e. where the sub-bunches are well separated and their distance, thus the THz emission frequency, can be tuned with the RF compressor injection phase, is the the so-called "deep" over-compression regime (Fig.9).

Interferometer measurements have been performed mainly in this regime. The comb current

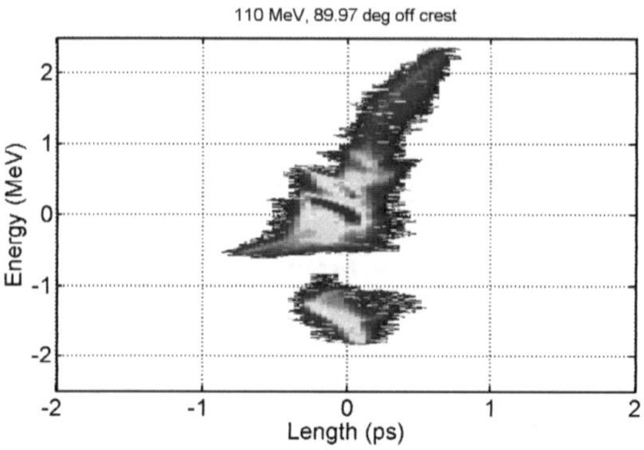

Figure 7. Measured longitudinal phase space for a four-pulses comb beam at the maximum compression phase; the title reports the average energy and the compression phase (200 pC total charge).

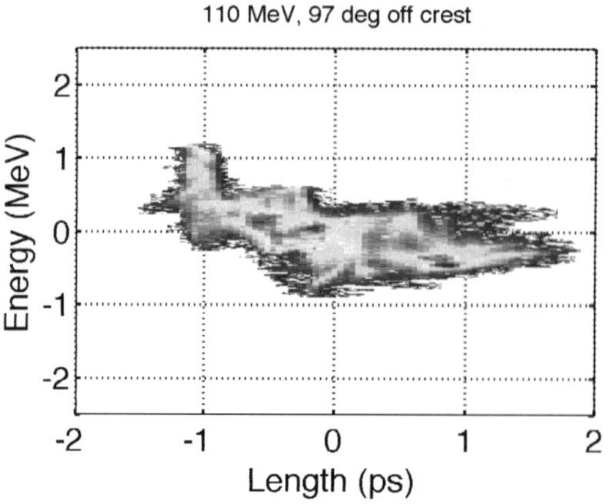

Figure 8. Measured longitudinal phase space for a four-pulses comb beam in the over-compression regime; the title reports the average energy and the compression phase (200 pC total charge).

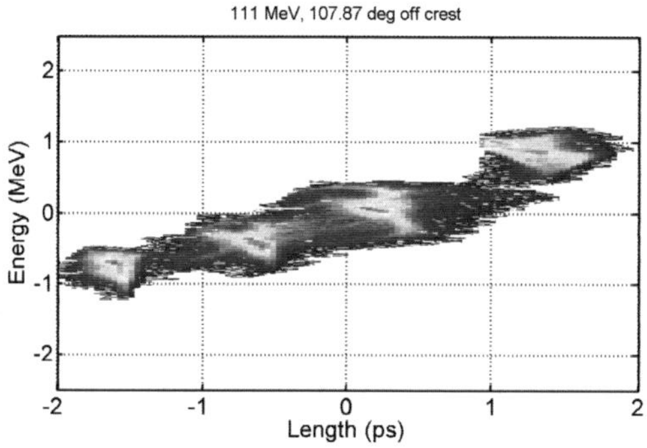

Figure 9. Measured longitudinal phase space for a four-pulses comb beam in the "deep" over-compression regime; the title reports the average energy and the compression phase (200 pC total charge).

profile, shown in Fig.10, has been measured at the beginning of the dogleg, i.e. after the first bending magnet in Fig.5. The four-pulses comb train has been then transported along the dogleg line down to the TR screen. The measured autocorrelation function of the modulated current profile shows clearly five (Fig.11, blue curve) out of the seven peaks expected from a four-pulses comb train (Fig.11, red curve) [18]. A reason for the absence of the outer peaks in the measured interferogram might be the effect due to the transport of the comb beam. Indeed, the longitudinal distribution emerging from the dogleg is affected by both the beam overall energy chirp (e.g. the chirp between sub-pulses) and the sub-pulses chirp (e.g. the chirp within a single sub-pulse). If the dogleg R_{56} parameter is not zero, two sub-pulses might happen to merge into a single one so that the emission will then be driven by a three-pulses comb beam.

In Fig.12 the form factor retrieved from the measured interferogram (blue curve) is compared to the one (red curve) retrieved from the current profile shown in Fig.10. The discrepancy in the emission peaks of the two curves can also be explained by considering the effect of the beam transport through the dogleg: under some conditions, it may act as a decompressing section,

Journal of Physics: Conference Series **359** (2012) 012018 doi:10.1088/1742-6596/359/1/012018

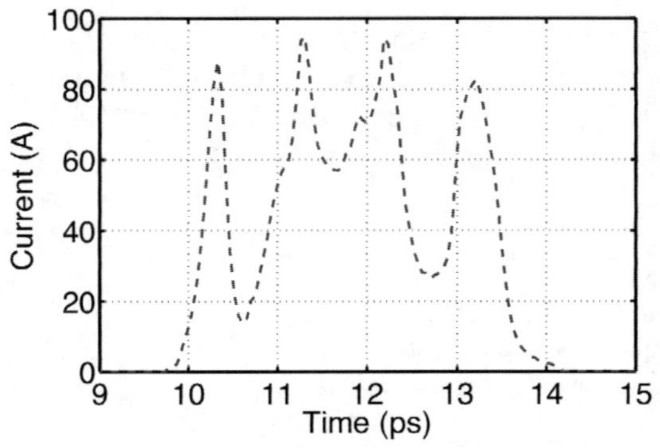

Figure 10. Current profile of a four-pulses comb beam as measured at the beginning of the dogleg.

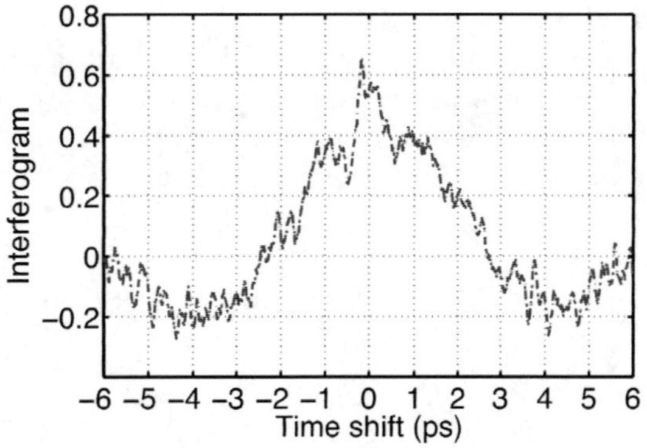

Figure 11. Measured interferogram (blue curve) and expected one from the modulated measured beam shown in Fig.10.

contributing to increase the pulse length and pulses separation, shifting the emission frequency in the sub-THz range.

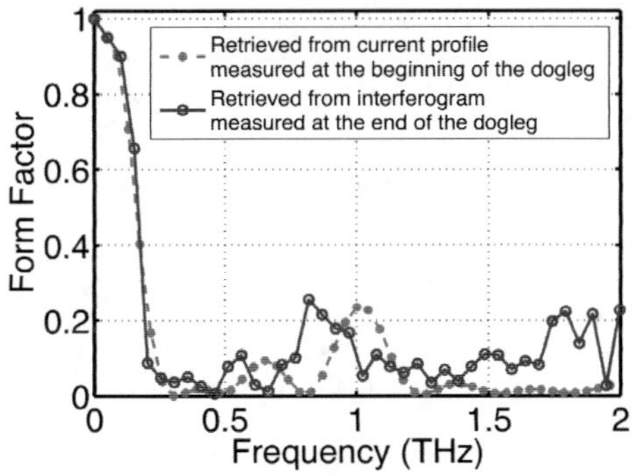

Figure 12. Form factor retrieved from current profile measured at the beginning of the dogleg (red curve), in comparison with the one retrieved by measuring the autocorrelation function of the CTR produced by the four-pulses comb beam.

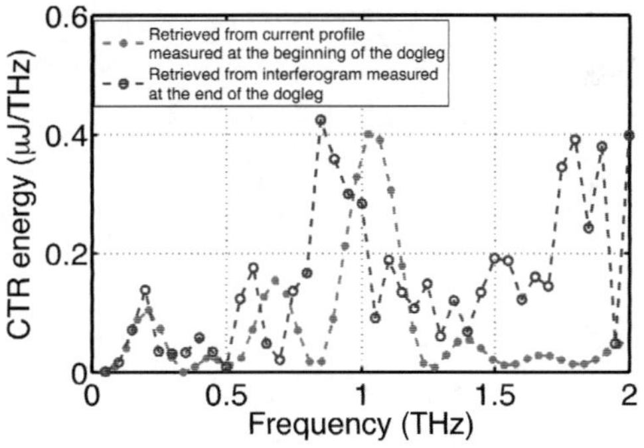

Figure 13. CTR energy per pulse from the measured interferogram (blue curve) and as expected (red curve) from the four-pulses comb beam measured at the beginning of the dogleg. Both curves are obtained by taking into account the single particle TR spectrum in far field approximation and for a finite target size. The experimental system transfer function is also considered.

The measured CTR energy per pulse has then been retrieved (Fig.13, blue curve) taking into account the single particle TR spectrum and the transfer function of the experimental system. A well defined narrow band peak centered around 0.8 THz is visible, as expected from the interferogram peaks distance.

Journal of Physics: Conference Series **359** (2012) 012018 doi:10.1088/1742-6596/359/1/012018

5. Conclusions

A linac-driven high peak power THz radiation has been realized and characterized at SPARC, taking advantage of the high brightness electron beams produced there. In particular, the possibility to generate and control high charge (200 pC), tailored comb trains makes SPARC unique in the survey of high peak power, narrow band, tunable THz radiation.

References

[1] L. Giannessi et al., Phys. Rev. Lett. **106**, 144801 (2011).
[2] L. Giannessi et al., Phys. Rev. Special Topics - Accel. and Beams **14**, 060712 (2011).
[3] M. Labat et al., Proceedings of FEL Conference 2011, WEOB3, Shangai (2011).
[4] D. Filippetto et al., Proceedings of FEL Conference 2011, WEOCI2, Shangai (2011).
[5] J.S.Nodvick and D.S. Saxon, Phys.Rev. **96**, 180 (1954).
[6] R. Lai and A.J. Sievers, Nucl. Instrum. and Meth. A **397**, 221-231 (1997).
[7] J. D. Jackson, Classical Electrodynamics, Third Edition, J.Wiley & Sons, Inc. (1995).
[8] M. Castellano et al., Nucl. Instrum. and Meth. A **435**, 297-307 (1999).
[9] S. Casalbuoni et al., PRST - AB **12**, 030705 (2009).
[10] L. Serafini and M. Ferrario, AIP Conf. Proceedings **581**, 87 (2001).
[11] M. Ferrario et al., Phys. Rev. Lett. **104**, 054801 (2010).
[12] J. G. Neumann et al., J. Appl. Phys. **105**, 053304 (2009).
[13] M. Boscolo et al., Nucl. Instrum. and Meth. A, **577**, 409-416 (2007).
[14] M. Ferrario et al., Nucl. Instrum. and Meth. A **637**, S43-S46 (2011).
[15] N. I. Landy, IEEE 1-4244-1449-0/07/$25.00 c, 1-2 (2007).
[16] D. H. Martin and E. Puplett, Infrared Physics, vol. 10, pp. 105-109 (1970).
[17] E. Chiadroni et al., Proceedings of IPAC 2010, TUOARA03, Kyoto (2010).
[18] E. Chiadroni et al., Proceedings of IPAC 2011, THPS101, San Sebastian (2011).

AUTHOR INDEX